THERAMIN 2020 Conference: Thermal Treatment of Radioactive Waste

IOP Conference Series: Materials Science and Engineering
Volume 818

Manchester, United Kingdom
4 - 5 February 2020

ISBN: 978-1-7138-2676-7
ISSN: 1757-8981

Printed from e-media with permission by:

Curran Associates, Inc.
57 Morehouse Lane
Red Hook, NY 12571

proceedings.com

Some format issues inherent in the e-media version may also appear in this print version.

Institute of Physics
Dirac House, Temple Back
Bristol BS1 6BE UK

Phone: 44 1 17 929 7481
Fax: 44 1 17 920 0979

techtracking@iop.org

Additional copies of this publication are available from:

Curran Associates, Inc.
57 Morehouse Lane
Red Hook, NY 12571 USA
Phone: 845-758-0400
Fax: 845-758-2633
Email: curran@proceedings.com
Web: www.proceedings.com

TABLE OF CONTENTS

IOP Conf. Series: Materials Science and Engineering **818** (2020) 011001 doi:10.1088/1757-899X/818/1/011001

THERAMIN Conference Proceedings: Preface

This Proceedings volume is dedicated to presenting the results of the EC THERAMIN (Thermal Treatment for Radioactive Waste Minimisation and Hazard Reduction) Project and other recent developments related to thermal treatment of radioactive waste.

The THERAMIN conference was held at the Mechanics Institute, Manchester on 4th and 5th February 2020. The conference brought together over 80 researchers and industrialists from 10 countries to share knowledge and learning from experience on the following topics, aligned with the main themes of the THERAMIN project (shown in Figure 1):

- Strategic review of radioactive wastes that could potentially be thermally treated and available thermal treatment technologies (Work Package 2).
- Demonstration of thermal treatment technologies for selected waste stream/technology combinations (Work Package 3).
- Characterisation and disposability of thermally treated products (Work Package 4)

The THERAMIN project also aimed to disseminate the knowledge gained through the project by inviting those not involved in the project to participate in this conference and by organising individual training placements (hosted by VTT, The University of Sheffield and CEA) and a technical training school for 20 early career researchers and students at the CEA's Le Visiatome Centre in Marcoule (Work Package 5).

THERAMIN aimed to identify which wastes could benefit from thermal treatment, which treatment technologies were under development in participating countries, and how these could be combined to deliver a wide range of benefits, including:

- Significant volume reduction.
- Waste passivation.
- Destruction of organic materials.

These benefits reduce risks during waste storage and support development of safety cases for geological disposal.

The conference was hosted by the University of Sheffield, Galson Sciences Limited and the National Nuclear Laboratory (UK partners in the Theramin project) and the HADES Facility at the University of Sheffield.

The conference organising committee consisted of:

Professor Neil Hyatt, Department of Materials Science & Engineering, The University of Sheffield.

Dr Jenny Kent, Galson Sciences Ltd, UK.

Professor Anthony Banford, National Nuclear Laboratory, UK.

The organising committee would like to thank the authors, speakers and participants for their contribution to the success of the THERAMIN conference and for supporting the peer review of papers submitted for inclusion within the conference proceedings. In addition, Dr Sally Scourfield (Galson Sciences Ltd. and Dr Laura Gardner (The University of Sheffield), played a pivotal role in the organisation of the meeting.

All papers published in this volume of *IOP Conference Series: Materials Science and Engineering* have been peer reviewed through processes administered by the proceedings Editors. Reviews were conducted by expert referees to the professional and scientific standards expected of a proceedings journal published by IOP Publishing.

Figure 1. Structure of and interactions between the work packages in the THERAMIN project.

The THERAMIN project was carried out by a consortium of 12 partners representing a European-wide community of experts on thermal treatment technologies and radioactive waste management and disposal. The project included an advisory group of waste owners / site operators and management organisations to provide an end-user view. This project has received funding from the Euratom research and training programme 2014-2018 under grant agreement No 755480. The work presented reflects the views of the authors and the European Commission is not responsible for any use that may be made of it. Finally, this meeting was held in association and supported the HADES / MIDAS Facility at The University of Sheffield, established with financial support from the UK Department for Business, Energy and Industrial Strategy and EPSRC under grant reference EP/T011424/1.

IOP Conf. Series: Materials Science and Engineering **818** (2020) 011002 doi:10.1088/1757-899X/818/1/011002

Peer review statement

All papers published in this volume of *IOP Conference Series: Materials Science and Engineering* have been peer reviewed through processes administered by the proceedings Editors. Reviews were conducted by expert referees to the professional and scientific standards expected of a proceedings journal published by IOP Publishing.

IOP Conf. Series: Materials Science and Engineering **818** (2020) 012001 doi:10.1088/1757-899X/818/1/012001

Thermal treatment for radioactive waste minimisation and hazard reduction: overview and summary of the EC THERAMIN project

S J Scourfield[1], J E Kent[1*], S M Wickham[1], M Nieminen[2], S Clarke[3], B Frasca[4]

[1] Galson Sciences Ltd. (UK), [2] VTT Technical Research Centre of Finland Ltd. (Finland), [3] National Nuclear Laboratory (UK), [4] Andra (France)

Email: jek@galson-sciences.co.uk

Abstract. The EC THERAMIN project aimed to identify which wastes could benefit from thermal treatment, which treatment technologies are under development in participating countries, and how these could be combined to deliver a wide range of benefits. Thermal pre-treatment or immobilisation processes result in significant volume reduction, waste passivation and destruction of organic materials, which reduces risks during waste storage and supports development of safety cases for geological disposal. This paper presents the key conclusions from the project. The potential for thermal treatment of European radioactive waste streams was evaluated and a "Value Assessment" framework was developed to assist in decision making, taking into account all stages of the waste management lifecycle. An overview is given of the strategy followed in performing demonstration trials and subsequent waste product characterisation for a range of waste groups. Case studies for select demonstrator-waste group combinations provide information about these processes in greater depth, including details about the treatment technique, resulting waste product and its characterisation, and disposability implications. Finally, the key conclusions from the project are summarised.

1. Introduction

The THErmal treatment for RAdioactive waste MINimisation and hazard reduction (THERAMIN) project was a European Commission (EC) programme of work jointly funded by the Horizon 2020 Euratom research and innovation programme and European nuclear waste management organisations (WMOs). THERAMIN ran from June 2017 to May 2020. Twelve European organisations (including WMOs, research institutes and consultancies), from seven European countries (Belgium, France, Germany, Lithuania, Slovakia, Finland and the UK), participated in THERAMIN.

The overall objective of the project was to provide improved safe long-term storage and disposal of intermediate-level wastes (ILW) and low-level wastes (LLW) suitable for thermal processing. Work carried out within the project aimed to identify radioactive wastes that could benefit from thermal treatment, which treatment technologies were under development in participating countries, and how these could be combined to deliver a wide range of benefits. The work programme provided a vehicle for co-ordinated EU-wide research and technology demonstration, and consisted of five Work Packages (WPs): WP1 involved project management and coordination; WP2 evaluated the potential for thermal treatment of waste streams from across Europe; in WP3, the application of selected thermal treatment technologies to radioactive waste management was demonstrated and evaluated; in WP4, the disposability of thermally treated radioactive waste products was assessed; and WP5 concerned the synthesis of project outcomes and their dissemination to interested parties. The links between these work packages are shown in Figure 1 and each is discussed in turn below.

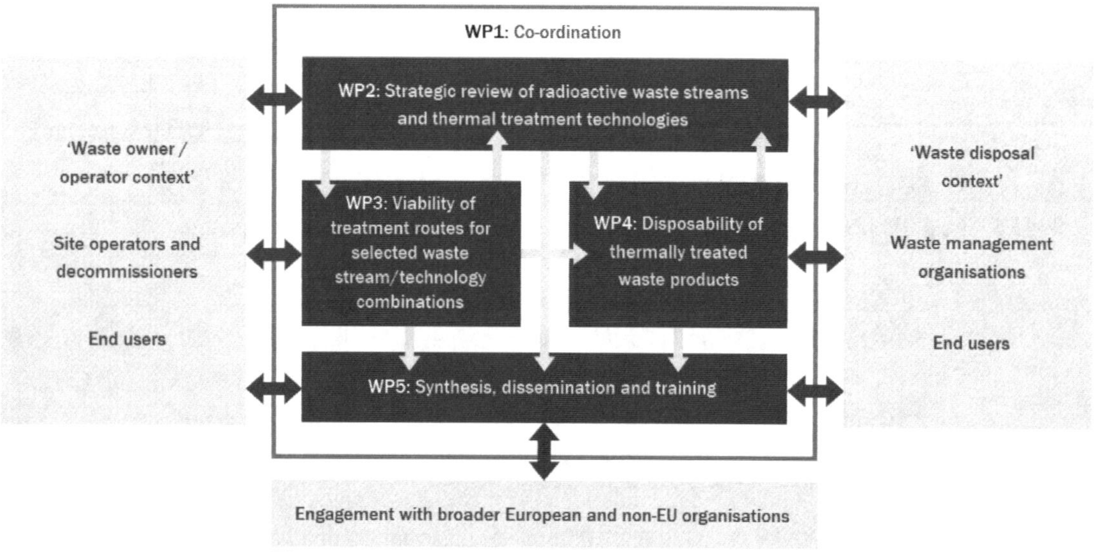

Figure 1. Structure of the THERAMIN project.

2. Work Package 2: Opportunities for thermal treatment: a strategic review

WP2 evaluated the potential for thermal treatment of particular waste streams across the EU and Ukraine, including consideration of where the greatest benefits from EU-wide collaboration may be realised, and development of a methodology to support decision-making on the potential use of thermal treatment technologies to treat various radioactive wastes [1]. To achieve this, wastes in Europe that are potentially suitable for thermal treatment, and potential thermal treatment technologies available in Europe, were identified. This information was used to derive a viability matrix in which the suitability of particular technologies to treat particular waste groups was shown [2].

Inventory information on radioactive waste streams that could benefit from thermal treatment, including any specific issues associated with treatment, processing, packaging or transport, was obtained by distributing a questionnaire to the seven THERAMIN-participating countries and Ukraine. The resulting database provided volumetric information about potential candidate wastes for thermal treatment; wastes occurring in several countries include alpha-contaminated waste, bitumen- or cement-conditioned solid waste, metallic waste, mixed waste, ion exchange materials and sludges/concentrates. It also enabled a strategic analysis of the benefits of thermal treatment for each country, by considering the drivers for thermal treatment and the risks and barriers to applying this treatment.

The availability and maturity of thermal treatment technologies within Europe was also summarised within WP2. The survey identified a wide range of techniques that could be grouped into three high-level process types: thermal treatment for volume reduction and passivation (eight technologies), conditioning by immobilisation in glass (eight technologies), and conditioning by immobilisation in ceramic or glass-ceramic matrices (one technology). Potential thermal treatment technologies and example facilities were identified, and the range of wastes they had demonstrably treated, or are theoretically capable of treating, was considered.

The final task of WP2 was to develop a Value Assessment methodology, which was intended to assist stakeholders in assessing the 'value' of a treatment technology when used to treat a particular radioactive waste stream. Value, in this context, was defined as realisable benefit in safety, monetary and environmental outcomes from implementing an option at a specified time. This included benefits and challenges across all stages of the waste management lifecycle. The assessment framework was based on a range of attributes that considered all of these different aspects of value across the whole waste management lifecycle [1].

The Value Assessment methodology was developed through the THERAMIN project, before being tested and refined at a workshop involving THERAMIN project Partners and End Users in the project's final year. The methodology is a generic starting point that can be tailored to the needs and context in which the assessment is being completed and the stage in the decision-making process that it is intended to inform.

3. Work package 3: Demonstration of thermal treatment technologies

Work within WP3 successfully demonstrated the thermal treatment of a range of waste groups in demonstrations carried out at existing thermal treatment facilities across the EU – namely, the SHIVA, In-can Melter, Geomelt®, thermal gasification, vitrification and hot isostatic pressing (HIP) demonstrations. These trials are described in detail in [3] and are summarised below. Photographs of selected pilot and full-scale plants are shown in Figure 2.

SHIVA (Advanced Incineration-Vitrification Hybrid System): the CEA incinerated waste using a plasma burner and vitrified the resulting ashes in a cold-wall direct glass induction melting system. The process is well suited to treating organic and mineral waste with high alpha contamination and results in a homogeneous vitrified product.

In-can Melter: the CEA vitrified inactive ash (by-products of incineration of organic waste) within a metallic crucible, which was heated in a refractory furnace using electrical resistors. This allowed in-container vitrification and the production of a crystallised glass product.

GeoMelt® in-container vitrification: the National Nuclear Laboratory (NNL) treated two combinations of surrogate wastes, with added radioactive tracers: cemented products (surrogate sea disposal drums) and soil; and Magnox sludge and clinoptilolite ion exchange media. Both experiments successfully demonstrated co-processing and resulted in a vitrified waste product.

Thermal gasification: VTT treated organic ion exchange resins, reducing the volume significantly and producing a fine dust product requiring further immobilisation e.g. by geopolymer encapsulation, before disposal. The method can also be used for low-level operational waste containing organic matter, if crushed before treatment.

Vitrification: inorganic liquid (chrompik) waste was vitrified by heating with glass frit in an inductively heated melting crucible by VUJE and JAVYS. Following the evaporation of water and continued heating for 6 hours, the resulting vitrified product was poured into a storage container. Off-gas from the process was decontaminated via a sorption column.

HIP: NNL and the University of Sheffield used HIP to demonstrate treatment of surrogates for Magnox sludge and clinoptilolite. The University of Sheffield completed an active trial, demonstrating the use of an active furnace isolation chamber (AFIC) that allows processing of radioactive waste simulants without risk of contamination to the processing equipment [4].

These processes and the resulting products are illustrated in Figure 2.

Figure 2. Thermal treatment facilities (from top left, clockwise: In-can melting pilot plant, CEA/Orano; VTT's pilot scale thermal gasification test facility; model of the vitrification facility, VUJE/Javys; Geomelt rig, NNL).

4. Work Package 4: Characterisation and disposability of thermally treated waste products

In order to be disposed of, radioactive waste products must comply with the Waste Acceptance Criteria (WAC) for a disposal facility. The WAC identify the characteristics required in a waste product in order to ensure that the waste cannot have a significant detrimental impact on the long-term safety provided by the disposal facility. Compliance with the WAC is confirmed by characterisation of the thermally treated products.

4.1. Identification and review of criteria and requirements for the disposability of thermally treated waste products

WP4 involved the identification and review of criteria and requirements for the disposability of thermally treated waste products; the characterisation of thermally treated waste products and secondary waste; and consideration of the downstream / safety case implications for disposability of the waste products. This also required consideration of the differing disposal concepts and national contexts of the countries participating in THERAMIN [5].

A set of generic disposability criteria were derived following review of national WAC and other disposability requirements applicable in individual countries (collectively referred to here as 'national disposability criteria') [5]. To this end, information relevant to the management of radioactive waste and application of thermal treatment, such as the approach to classifying radioactive waste, a summary of the inventory, the status of disposal facilities, planned development activities and known issues, was obtained from each of the seven countries participating in THERAMIN.

The generic disposability criteria were intended to be applicable to any packaging or disposal concept, any thermally treated waste, and any disposal environment. These criteria were [6]: no free liquid or gas; permeability and/diffusivity of the waste; no or limited hazardous material content; immobilisation of radionuclides; limited voids / limited porosity; homogeneity of the thermally treated product; leaching behaviour of the waste product; mechanical resistance of the waste product; no metal with a redox lower than 0.84 V Standard Hydrogen Electrode (SHE); and the thermal behaviour of the waste.

The criteria were also designed to highlight the impact thermal treatment can have on waste disposability. Characterisation requirements for thermally treated products in order to test compliance against each criteria, and suitable analytical techniques that could enable this, were then identified [7].

4.2. Characterisation of thermally treated waste products

Characterisation tests were completed for samples of thermally treated products from all the demonstrations carried out within WP3 (Section 3) and some additional samples from thermal treatment processes not tested in the project (shown in Table 1). A common basis for solid characterisations was chosen to test [7]:

- the degree of homogeneity of the sample and to verify the absence of free liquid or gas;
- the overall chemical composition of a homogeneous sample of the local compositions of a heterogeneous sample;
- and the amorphous or crystalline nature of a sample and the structure of the crystals present in the crystallised sample.

The analytical techniques available to all THERAMIN partner laboratories which constitute this common basis were scanning electron microscopy, X-ray fluorescence spectrometry, electron microprobe, inductively coupled plasma analysis after dissolution of the solid and X-ray diffraction. The chemical durability of the samples was estimated by leaching tests based on the ASTM Standard Test Method C 1285 – 14 "Standard Test Methods for Determining Chemical Durability of Nuclear, Hazardous, and Mixed Waste Glasses and Multiphase Glass Ceramics: The Product Consistency Test (PCT)" [8]. Additionally, depending on the nature of the samples and the national waste management context, other techniques, such as total organic and inorganic carbon analyses, gas physisorption to determine the specific surface area of a powder sample, thermal conductivity or transmission electron microscopy were used if available.

The characterisation results for 17 samples, including glass, ceramic, ashes and geopolymerised products following thermal treatment are presented in [7], including a description of the sample macrostructure and microstructure, chemical composition and chemical durability.

IOP Conf. Series: Materials Science and Engineering **818** (2020) 012001 doi:10.1088/1757-899X/818/1/012001

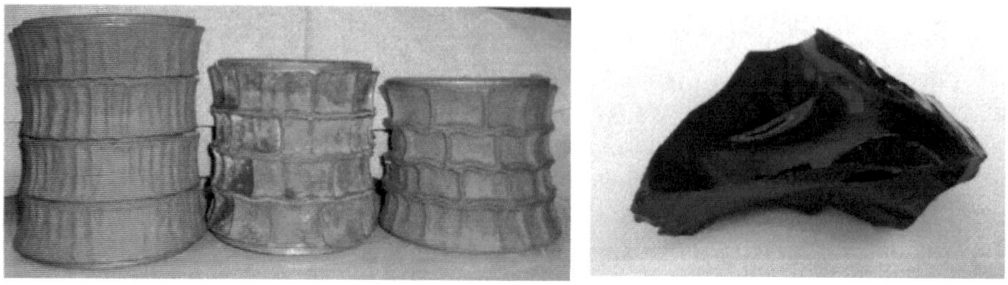

Figure 3. Thermal treatment products (left, HIP containers, NNL; right, glass sample from the SHIVA trial, CEA/Orano).

4.3. Disposability and safety case implications

The disposability of thermally treated waste products and the resulting secondary wastes was evaluated for a subset of the characterised samples, based on the disposability criteria of the country in question or the generic WAC derived in [6]. These assessments have been completed for both near-surface and geological disposal facilities by the WMOs within the THERAMIN End User Group.

As an example, the disposability of ion exchange resins that have been incinerated and vitrified using the SHIVA process (as described in Section 3) has been qualitatively evaluated by Andra [9]. Based on characterisation performed by CEA [7], the final product consists of an amorphous alumina-borosilicate glass with radionuclide incorporation in the glass matrix. The characterisation results indicate that organic matter has been removed by thermal treatment, leading to favourable characteristics for geological disposal due to the absence of organic complexing compounds and potential sources of hydrogen gas release. The rate of radionuclide release will be directly dependent on the chemical durability of the glass product; the long-term behaviour of the final product needs to be assessed in more detail but is considered to compare favourably to the instant release model associated with the baseline cemented ion exchange resin wasteform.

5. WP5: Dissemination of project results and outcomes

WP5 aimed to disseminate knowledge and outcomes of the project within the technical community and more widely. This was achieved through a variety of activities, including training, organisation of a project conference and development of a project synthesis report.

Training placements were hosted by the University of Sheffield, VTT and the CEA in conjunction with the thermal treatment trials conducted as part of WP3. These placements were attended by students and researchers either early in their career or from countries with less well-developed thermal treatment technologies. A technical training school, hosted by the CEA, was held in Marcoule in June 2019. The course, which included site visits to the CEA's thermal treatment rig hall and the neighbouring EDF-Cyclife incinerator and metal melting facility, was attended by 20 people. Presentations from the course, which aimed to share the key learning points of the project with a wider audience of non-project participants, were also made available on the THERAMIN website.

In February 2020, the THERAMIN Conference was held in Manchester which brought together project participants, researchers and WMO representatives worldwide, and for which this publication constitutes the conference proceedings. The EC also funded attendance at the conference for three students, as part of the training activities described above.

Finally, WP5 will deliver a project synthesis report, which will be publicly available from end May 2020 and will contain an overview of the main research, activities and outcomes of the project.

6. Conclusions from the THERAMIN project

The THERAMIN project successfully demonstrated the applicability of six different thermal treatment technologies to a range of waste groups (labelled WP3 in Table 1) and provided tools to compare these technologies with each other and with established baselines, where present. The thermally treated products of these and other trials have also been characterised (labelled WP4 in Table 1) and these data used to undertake preliminary disposability assessments. Although further studies of long-term leaching behaviour are required, the destruction of organic species by thermal treatment has been demonstrated to provide benefits in terms of reduced waste volume, reduced gas generation and removal of complexants that could increase the rate of radionuclide transport within the disposal facility.

A community of thermal treatment specialists has been developed through the THERAMIN project, which provides a forum for sharing experience, understanding challenges and discussion and identification of potential solutions. At the THERAMIN conference, opportunities to continue to develop this community were identified, including related work in the new EC PREDIS (Predisposal waste management) project and annual meetings in the UK supported by the University of Sheffield HADES facility [10].

Table 1. Waste-technology combinations tested in THERAMIN WP3 [3] and WP4 [7].

Technology	Sludge	Cement conditioned wastes	Organic ion-exchange material	Inorganic ion-exchange material	Ash	Inorganic liquor	Mixed solid waste	Uranium
GeoMelt®	WP3 WP4	WP3 WP4		WP4				
Hot Isostatic Pressing	WP3 WP4			WP3 WP4				WP4
SHIVA			WP3 WP4					
In-can melter					WP3 WP4			
VICHR Vitrification						WP3 WP4		
Thermal Gasification			WP3 WP4					
Plasma vitrification		WP4	WP4				WP4	
Incineration					WP4		WP4	

Acknowledgments

This project has received funding from the Euratom research and training programme 2014–2018 under grant agreement No. 755480. This paper reflects only the authors' views, and the European Commission is not responsible for any use that may be made of it. The authors are grateful to the THERAMIN Project Partners who contributed to the deliverables summarised and referenced herein, including the following organisations:

- VTT Technical Research Centre of Finland (VTT), Finland
- Agence Nationale pour la Gestion des Déchets Radioactifs (Andra), France
- Orano, France
- Commissariat à l'Énergie Atomique et aux Énergies Alternatives (CEA), France
- Galson Sciences Limited (GSL), United Kingdom
- Forschungszentrum Jülich GmbH (FZJ), Germany

- Lithuanian Energy Institute (LEI), Lithuania
- National Nuclear Laboratory Limited (NNL), United Kingdom
- Belgian Agency for Radioactive Waste and Enriched Fissile Materials (ONDRAF/NIRAS), Belgium
- Studiecentrum voor Kernenergie/Centre d'Etude de l'Energie Nucléaire (SCK•CEN), Belgium
- The University of Sheffield (USFD), United Kingdom
- VUJE a.s. (VUJE), Slovakia

References

[1] Fuller A, Doudou S, Kent J, Harvey E, Wickham S 2020 Assessing the value of thermal treatment technologies, *IOP Conf. Ser.: Mater. Sci. Eng* same issue

[2] Doudou S et al. 2020 Strategic Study of Thermal Treatment of European Radioactive Wastes *IOP Conf. Ser.: Mater. Sci. Eng* same issue

[3] NNL et al. 2019 *Summary of Demonstration Trials Carried Out Under WP3*, European Commission THERAMIN Deliverable D3.2, November 2019

[4] Gardner L J, Walling S A and Hyatt N C 2020 Hot isostatic pressing: thermal treatment trials of simulant inactive and radioactive UK ILW *IOP Conf. Ser.: Mater. Sci. Eng* same issue

[5] Harvey L, Galson D, Catherin S, Romero M-A, Fournier M, Fuller A and Wickham S 2020 Development of *Generic criteria for evaluating the disposability of thermally treated wastes, IOP Conf. Ser.: Mater. Sci. Eng* same issue

[6] Andra et al. 2018 *Waste Acceptance Criteria and requirements in terms of characterisation*, European Commission THERAMIN Deliverable D4.1, May 2018.

[7] Andra et al. 2019 *Characterization of thermally treated waste products*, European Commission THERAMIN Deliverable D4.2, July 2019.

[8] ASTM International 2014 *Standard test method for determining chemical durability of nuclear, hazardous and mixed waste glasses and multiphase glass ceramics: the product consistency test (PCT)*. ASTM Standard C1285/14, West Conshohocken, PA, United States

[9] Frasca B, Griffault L, LeFevre J, Martin C and Romero M-A 2020 *Influence of thermal treatment on the disposability of spent ion exchange resins in deep geological repository: a French case, IOP Conf. Ser.: Mater. Sci. Eng* same issue.

[10] Hyatt N C, Corkhill C L, Stennett M C, Hand R J, and Thorpe C L 2020 *The HADES Facility for High Activity Decommissioning Engineering & Science: part of the UK National Nuclear User Facility, IOP Conf. Ser.: Mater. Sci. Eng* same issue

Strategic Study of Thermal Treatment of European Radioactive Wastes

S Doudou[1*], E K Swain-Phipps[1], A J Fuller[1], S M Wickham[1], N Daniels[2], E Fourcy[3], M Fournier[4], B Frasca[5], K Meert[6], A Mikusova[7], M Olin[8], R Poskas[9], C Scales[10]

[1] Galson Sciences Ltd (UK), [2] Juelich (Germany), [3] Orano (France), [4] CEA (France), [5] Andra (France), [6] ONDRAF/NIRAS (Belgium), [7] VUJE (Slovakia), [8] VTT (Finland), [9] LEI (Lithuania), [10] NNL (UK)

*sd@galson-sciences.co.uk

Abstract. Under Work Package 2 (WP2) of THERAMIN, strategic reviews of thermal treatment technologies and European radioactive waste streams for which thermal treatment could bring benefits have been undertaken. This paper presents the outcomes of an international review of inventory information concerning European wastes potentially suitable for thermal treatment, a review of thermal treatment technologies, and a strategic gap analysis to identify countries where there are significant waste arisings with potential benefits for thermal treatment, but little prospect of these countries developing thermal treatment facilities independently.

1. Introduction

The overall objective of THERAMIN (The **The**rmal treatment for **ra**dioactive waste **min**imisation and hazard reduction) is to identify improved options for safe long-term storage and disposal of intermediate-level wastes (ILW) and low-level wastes (LLW) suitable for thermal processing. The work programme provides a vehicle for coordinated EU-wide research and technology demonstration. It is designed to provide improved understanding and optimisation of the application of thermal treatment in radioactive waste management programmes across Europe, and to move thermal technologies higher up the Technology Readiness Level (TRL) scale.

Under Work Package 2 (WP2) of THERAMIN, strategic reviews of thermal treatment technologies and European radioactive waste streams for which thermal treatment could bring benefits have been undertaken. WP2 is structured in five tasks:

- Task 2.1: development of a European database of radioactive waste groups suitable for thermal treatment.
- Task 2.2: implementing a strategic analysis of the benefits of thermal treatment of the waste groups identified and the risks and barriers that exist to such treatment.
- Task 2.3: summarising the availability and maturity of different thermal treatment technologies.
- Task 2.4: constructing a summary matrix of waste groups and applicable thermal treatment technologies.

- Task 2.5: undertaking a value assessment of the strengths and challenges of selected waste group / thermal treatment combinations.

This paper presents the outcomes of Tasks 2.1 to 2.4. Specifically, this paper focuses on the outputs of an international review of inventory information concerning European wastes potentially suitable for thermal treatment, a review of thermal treatment technologies, and a strategic gap analysis to identify countries where there are significant waste arisings with potential benefits for thermal treatment, but little prospect of these countries developing thermal treatment facilities independently.

2. European wastes Potentially Suitable for Thermal Treatment

A Europe-wide review was undertaken to gather inventory information on radioactive waste streams that could benefit from thermal treatment. Inventory information was gathered by means of a questionnaire from THERAMIN Partner and End User Group countries (Belgium, Finland, France, Germany, Lithuania, Slovakia, Switzerland, UK). In addition, the European Commission (EC) encouraged inclusion of Ukraine, and its inventory was also captured in the waste database. Waste information from the EC project Microbiology In Nuclear waste Disposal (MIND) was also sought, which resulted in the addition of the following countries: the Czech Republic, the Netherlands, Spain, and Sweden.

The waste streams identified were categorised according to waste type and composition into 14 generic waste groups (Table 1) and compiled in an electronic database for presentation and data analysis (Task 2.1, [1]). Each waste group may have specific issues associated with treatment, processing, packaging or transport, although for some groups the issues are common. The categorisation into generic groups was based on commonalities in the waste stream properties and composition. There are common waste types that occur in several countries, such as sludges, ion exchange materials, cement-conditioned wastes, and bitumen-conditioned waste, whilst others were identified as present in fewer countries (e.g. polymer-conditioned waste and non-organic liquid wastes).

The database was used to generate volume information for the generic waste groups across in the countries reviewed. Figure 1 provides a summary of the breakdown of wastes (both existing and future arisings) covered in the database, excluding Ukraine, which would dominate if included because its inventory, which represents the national inventory, is volumetrically much larger than those of other countries. It is noted that the estimated volumes of waste are those that have been deemed potentially suitable for thermal treatment by country contacts filling out the questionnaires. They do not represent the total inventories in the listed countries. In addition, the listed countries in Figure 1 are those that have identified a particular waste as a potential candidate for thermal treatment, noting that thermal treatment may be an alternative to other treatment options rather than an optimized solution. Other countries not listed may hold similar waste but manage it via an already established route.

The database was also used to enable a strategic analysis of the benefits of thermal treatment. The analysis considered the drivers for thermal treatment, and the risks and barriers to applying this treatment. For each country, the analysis described the risks if the wastes are left untreated and potentially without a management route, and identified those wastes for which thermal treatment is unlikely to bring significant benefits. This analysis was supported by a brief description of the national context and programme status in each participating country. The database and strategic analysis are documented in a report (Task 2.2, [2]).

Table 1. Generic waste groups defined in [2].

Generic waste group	Description
Alpha waste (including PCM)	Material contaminated with alpha-emitting radionuclides (e.g., plutonium, uranium, etc.). This waste includes PCM.
Bitumen-conditioned waste	Wastes that have been conditioned in a bitumen matrix. The nature of the original raw waste is varied.
Cement-conditioned solid waste	Wastes that have been conditioned in a cementitious matrix. The nature of the original raw waste is varied.
Filters	Filters are used to remove radionuclides and particulates from contaminated air or other media. Example filters include: HEPA, charcoal filters, and cartridge filters used to remove radionuclides and particulates from active effluent.
Graphite	Waste graphite from decommissioning of reactors that used graphite as part of the reactor design. This could include core graphite or graphite debris from the fuel assemblies.
Hazardous or Chemotoxic waste	Wastes which have chemotoxic properties (e.g., Be, Cd, Hg) or which are hazardous (e.g., asbestos).
Inorganic ion exchange material	Ion exchange materials used for the removal of soluble radionuclides (e.g., caesium) from liquid waste (e.g., irradiated fuel cooling pond water). Example inorganic resins include: zeolites, Ionsiv® and clays.
Metallic waste (pure or high content)	Waste containing pure metal or metal mixed with other materials.
Miscellaneous contaminated solid waste (including PVC)	Other miscellaneous solid waste that is non-metallic, e.g., maintenance wastes, decommissioning wastes, contaminated gravel, concrete, PVC, etc.
Polymer-conditioned waste	Wastes that have been conditioned in a polymer matrix. The nature of the original raw waste is varied.
Organic ion exchange material	Ion exchange materials composed of high-molecular-weight polymers. They are also used for the removal of soluble radionuclides from solution.
Other liquid waste (e.g. Chrompik)	Contaminated aqueous liquids which do not contain organics.
Organic liquids and oils	Contaminated liquid waste which contains organics such as oils or solvents.
Sludge and concentrates	Includes bulk sludge, residuals, and concentrates. Sludges arise in tanks, sumps and ponds, and comprise a mixture of particulate materials and water.

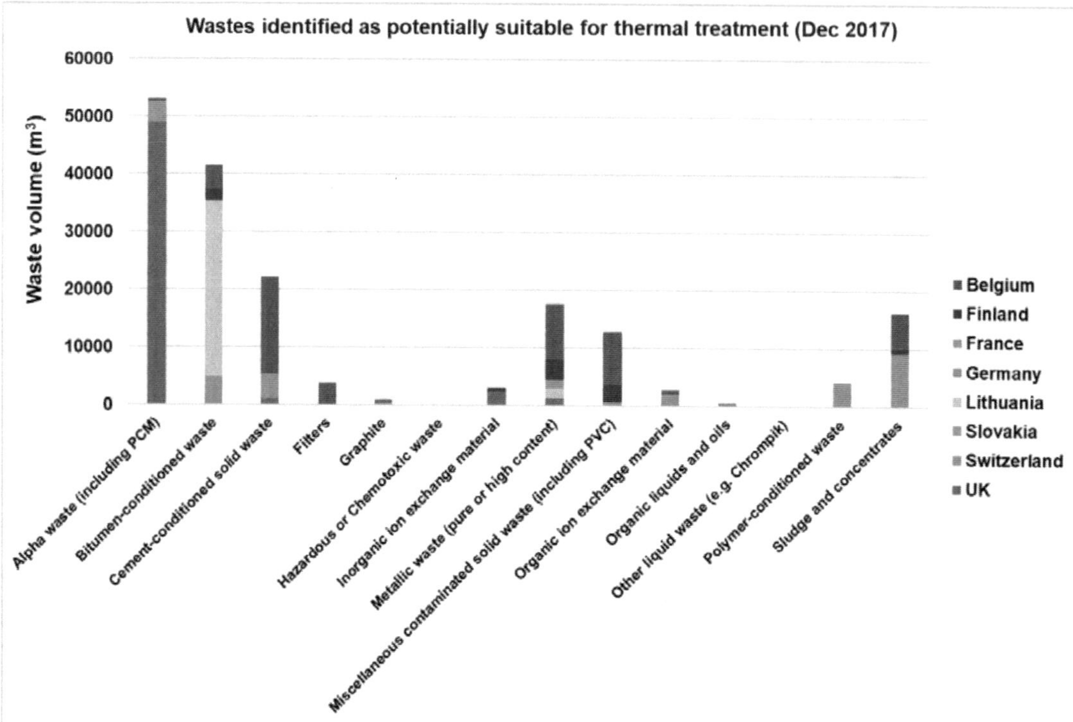

Figure 1. Waste volumes (existing and future arisings) identified for thermal treatment, grouped by generic waste group (excluding Ukrainian wastes). Note that the estimates provided correspond to waste streams identified as potentially suitable for thermal treatment and do not represent national inventories.

3. Thermal Technologies in Europe

Under Task 2.3, a report was produced describing potential thermal technologies available in Europe, and considering, at a high level, the range of wastes and materials that they have demonstrably treated, or are theoretically capable of treating [3]. The objective of Task 2.3 was to summarise the availability and maturity of thermal treatment technologies in Europe by:

- Providing a summary of the current status of each thermal technology with regard to treatment and processing of radioactive wastes.
- Evaluating the availability of thermal technologies to countries with significant waste arisings that could benefit from thermal treatment and processing.
- Collecting information to support production of a viability matrix and value assessment.

At a high level, the thermal technologies were first classified into three ***process categories*** according to the type of product produced as follows:

- Thermal processes that are expected to generate a product that requires additional conditioning (e.g., encapsulation) to allow it to become suitable for disposal. These technologies can be described as treatment processes that are employed to reduce chemical reactivity and/or potentially reduce volume.
- Thermal processes involving immobilisation of radioactive waste by encapsulation in a glass matrix or by direct incorporation into the glass matrix.
- Thermal processes involving immobilisation of radioactive waste by incorporation in a ceramic matrix or in a glass/ceramic composite matrix.

For each process category above, a sub-level of classification was used to sort existing technologies by heating mode and process wall material (e.g. refractory or cold wall, etc.). For each technology,

IOP Conf. Series: Materials Science and Engineering **818** (2020) 012002 doi:10.1088/1757-899X/818/1/012002

examples of existing facilities (commercial or laboratory scale) in Europe were then provided. The example facilities were only used to illustrate the possible variations that may exist within a treatment technology, and therefore are not exhaustive. Table 2 provides a summary of the classification that was adopted.

Table 2. Classification of thermal treatment and processing technologies considered in THERAMIN.

High-level process	Technology	Facility	Country
Treatment for volume reduction and passivation	Incineration with burner and refractory walls	JÜV 50/2 - Jülich JEN	Germany
		KTE incinerator	Germany
		EDF CENTRACO	France
	Rotary kiln incineration	IRIS	France
	Pyrolysis	Belgoprocess	Belgium
	Thermal gasification	VTT gasification	Finland
	Calcination	Widely used	France, UK
	Underwater plasma incineration	ELIPSE	France
	Hydrothermal Oxidation (HTO)	DELOS	France
	Induction metal melter	CARLA	Germany
		EDF CENTRACO	France
		Cyclife (formerly Studsvik)	Sweden
Conditioning by immobilisation in glass	Joule-Heated In-Can Vitrification	In-Can Melter and DEM & MELT (metallic inner wall), CEA	France
		GeoMelt (ceramic inner wall), NNL	UK
	Joule-Heated Ceramic Melter (JHCM)	VEK, PAMELA (both decommissioned)	Germany, Belgium
	Cold crucible induction melter (CCIM)	La Hague CCIM and Marcoule CCIM pilot	France
	Advanced CCIM (A-CCIM)	Marcoule A-CCIM pilot	France
	Indirect induction (metallic wall - hot metal pot)	VICHR	Slovakia
		La Hague and Sellafield	France, UK

IOP Conf. Series: Materials Science and Engineering **818** (2020) 012002 doi:10.1088/1757-899X/818/1/012002

High-level process	**Technology**	**Facility**	**Country**
	Coupled cold wall direct metal induction melting and plasma burner	PIVIC	France
	Coupled cold wall direct glass induction melting and plasma burner	SHIVA	France
	Refractory wall plasma burning and melting	Retech (ZWILAG)	Switzerland
		EUROPLASMA – Belgoprocess	Bulgaria
		Tetronics	UK
Conditioning by immobilisation in ceramic or glass-ceramic	HIP	NNL – Workington and University of Sheffield	UK

A process summary table was generated for each technology in order to provide information to support Tasks 2.4 (viability matrix) and 2.5 (value assessment). For each of the European facilities included in the technology survey, information was provided by THERAMIN partners on the following topics:

- Waste compatibility with technology (solid, liquid, organic, inorganic, metal, level of activity, etc.).
- Mode of heating.
- Containment.
- Nature of product.
- Limitations of technology.
- Maximum volumetric throughput.
- Continuous or batch process.
- Secondary waste generation.
- Maturity of technology (TRL).
- Target date for active commissioning of full-scale industrial facility.
- Technological complexity.
- Flexibility to treat a wide range of waste types.
- For specific facilities, the size of furnace or crucible (or volume of waste that can be treated in one batch).
- Scaleability.
- Investment and operational costs.

The information collected under Task 2.1 (wastes) and Task 2.3 (technologies) was then used to derive a viability matrix in which the suitability of particular technologies to treat particular waste types was shown. The waste types were organised in the same way as in the THERAMIN waste database so that they could be linked directly to specific countries and, if needed, to specific waste streams.

For each waste type – technology combination, the key factors determining whether the combination is viable or not were considered. It was proposed that viability should be considered in two distinct ways:

- Wastes that can be processed in principle by a particular technology.
- Wastes that have been demonstrably processed by a particular technology.

To assess the potential to apply each of the thermal treatment technologies to the identified waste groups, it was essential to review some of their key properties. The D2.3 report [3] focuses on a number of specific attributes. Firstly, the types of waste that could be treated by each technology were assessed. This focused specifically on whether the technologies can treat solids, liquids, or both solids and liquids. It also considered whether they could treat organic or metallic wastes. In addition, the maximum accepted levels of activity which each facility can accept were highlighted. Finally, the flexibility of each technology was also determined to assess whether it could be used to treat a wide range of wastes, and/or highly heterogeneous waste streams.

To determine accurately the viability of each treatment technology, it is important not just to assess their beneficial attributes, but also the potential limitations that may render treatment challenging or impossible. The technical limitations inherent in each technology were therefore also considered (e.g., some technologies might be challenged by significant metal fraction in the waste feed). Additionally, each of the facilities are bound by logistical constraints including their maximum treatment capacity and throughput, and whether they have been made to operate at an industrial scale. These aspects were also considered as they are likely to inform decision making (e.g., a low capacity method may not be ideal for treating large waste volumes). Finally, the TRL of the technique was considered as some technologies, although theoretically able to treat a waste with the identified properties, may still only be at the experimental or pilot stage.

Considering all of the above information, the D2.3 report [3] presented a viability matrix linking the generic waste groups to the thermal treatment technologies. Each technology was categorised as either having already been tested to treat that waste, potentially having applicability, having only limited applicability, or not being applicable. A number of the technologies have been tested in THERAMIN WP3 on a range of wastes (Table 3).

Table 3. Waste-technology combinations tested in THERAMIN WP3 (cells highlighted and ticked).

Technology	Sludge	Cement conditioned wastes	Organic ion-exchange material	Inorganic ion-exchange material	Ash	Inorganic liquor
GeoMelt	✓	✓				
Hot Isostatic Pressing	✓			✓		
SHIVA			✓	✓		
In-can melter					✓	
VICHR Vitrification						✓
Thermal Gasification			✓			

4. Gap Analysis

The viability matrix and the THERAMIN waste database were used to facilitate a strategic gap analysis to identify countries where there are significant waste arisings with potential to benefit from thermal treatment using technologies available within the country or in other European countries. Where gaps in domestic technologies and facilities were identified, suggestions were made of resources available to process the wastes in other countries.

It is noted that the gap analysis is based on the viability matrix, where the treatment technology applicability to specific generic waste types takes into account the properties of the wastes and the technical aspects of the treatment technology. It does not account for non-technical constraints and limitations on waste treatment such as constraints on moving wastes across international borders for treatment, regulatory barriers, or stakeholder implications. In addition, the mapping does not consider detailed characterisation data and specific properties of the waste streams within a generic waste group. Therefore, the information in the strategic analysis provides only a preliminary input and a starting point to aid decision making, rather than a definitive or optimised options appraisal for treating a particular waste stream.

The analysis concluded that although a few European countries may have the resources to thermally treat their own wastes, many other countries could benefit from cross-country collaboration and treatment of wastes outside their borders. This could provide a cost-effective option to treatment of challenging and problematic wastes. Further information and country-by-country analysis are provided in the D2.3 report [3].

5. Conclusions

This paper presents the outcomes of an international review of inventory information concerning European wastes potentially suitable for thermal treatment, a review of thermal treatment technologies available in Europe, and a strategic gap analysis to identify countries where there are significant waste arisings with potential benefits for thermal treatment, but little prospect of these countries developing thermal treatment facilities independently.

The waste streams identified by project partners were categorised into 14 different generic waste groups and included in an electronic database. The survey of European thermal treatment technologies identified a wide range of techniques that could be grouped into three high-level process types: thermal treatment for volume reduction and passivation (eight technologies), conditioning by immobilisation in glass (eight technologies), and conditioning by immobilisation in ceramic or glass-ceramic matrices (one technology). Both the inventory database and the thermal technology survey outcomes allowed the creation of a viability matrix which maps the generic waste groups to thermal treatment technologies that could potentially treat the identified wastes. All these tools (i.e., the waste database, technology survey, and viability matrix) can be used by THERAMIN project organisations to aid in their decision making regarding potential suitable thermal technologies for their wastes.

Finally, a strategic gap analysis was undertaken to identify those countries with wastes identified as suitable for thermal treatment but with no suitable domestic thermal treatment facility. Facilities with the technological capability to treat those wastes in other European countries were identified. Although this analysis does not account for non-technical limitations on movement of waste across international borders, it provides preliminary inputs for future decision making and further analysis taking into account wider considerations on specific waste characteristics and other facility constraints.

Acknowledgements

This project has received funding from the Euratom research and training programme 2014-2018 under grant agreement No 755480. This paper reflects the authors' views and the European Commission is not responsible for any use that may be made of it.

IOP Conf. Series: Materials Science and Engineering **818** (2020) 012002 doi:10.1088/1757-899X/818/1/012002

References

[1] Doudou S and Phipps E 2018 *Database of European radioactive wastes suitable for thermal treatment*, European Commission THERAMIN Deliverable D2.1.
[2] Doudou S *et al.* 2018 *European radioactive wastes suitable for thermal treatment*, European Commission THERAMIN Deliverable D2.2.
[3] GSL 2018 *Current Status of Thermal Treatment of Radioactive Waste Streams in the European Union*, European Commission THERAMIN Deliverable D2.3.

IOP Conf. Series: Materials Science and Engineering **818** (2020) 012003 doi:10.1088/1757-899X/818/1/012003

Assessing the value of thermal treatment technologies

A J Fuller, S Doudou, J E Kent, E J Harvey and S M Wickham
Galson Sciences Ltd, 5 Grosvenor House, Melton Road, Oakham, Rutland, LE15 6AX United Kingdom

ajf@galson-sciences.co.uk

Abstract. Work Package 2 of the THERAMIN project provides a framework to support decision making on the potential use of thermal technologies to treat various radioactive wastes. To do this, a methodology based on multi-attribute decision analysis has been devised to identify the benefits of applying thermal treatment techniques to a range of radioactive waste streams. The value assessment methodology used in THERAMIN follows an attribute-based approach with the potential benefits of the treatment technology assessed against different criteria, considering performance over the whole waste management lifecycle, from pre-treatment through to final disposal. The attribute groups considered are: operational safety, environmental impact, impact on disposability/long term safety, implementation, timescale, technical readiness, and cost impact. To guide the decision maker, each attribute in split into data categories (sub-criteria) and a summary of topics to be considered under each attribute is provided. The assessment attributes have been compiled by drawing on knowledge from across the THERAMIN partner organisations to make them as comprehensive as possible. Consideration is given to the whole of the waste lifecycle and to different European regulatory contexts.

1. Introduction to value assessment

The thermal treatment for radioactive waste minimisation and hazard reduction (THERAMIN) project is a European Commission (EC) programme of work jointly funded by the Horizon 2020 Euratom research and innovation programme and European nuclear waste management organisations (WMOs). Within the wider project, work package 2 (WP2) has undertaken a strategic review of radioactive waste streams, which could potentially benefit from thermal treatment. This work included assessing the value of thermal technologies when used to treat various waste streams.

The THERAMIN value assessment exercise builds on the UK Nuclear Decommissioning Authority's (NDA) value framework [1] (and other similar studies [2]) and is designed to provide a structured methodology to assess the 'value' of a chosen waste treatment technology when used to treat a waste stream of interest. Value, in this context, is defined as realisable benefit in safety, monetary and environmental outcomes from implementing an option at a specified time. This includes benefits and challenges across all stages of the waste management lifecycle. However, it should be recognised that value varies somewhat between stakeholders, with different people assigning greater or lesser importance to different attributes. Therefore, a value assessment, like the one outlined in this report, is a multi-attribute assessment that considers all of these different aspects of value across the whole lifecycle of the waste, and the treatment facility.

Value assessment exercises provide most benefit when conducted as workshops, engaging with a wide range of stakeholder who can bring knowledge and expertise from across the whole waste lifecycle. Alternatively, targeted assessments could also be conducted to focus on only a few key aspects of the technology that are of importance to a specific stakeholder or needed to answer a given question.

This paper was developed based on the outputs of a value assessment workshop attended by the THERAMIN partners, where detailed discussions focused on the approach that should be taken to assess the value of thermal treatment technologies, particularly those trialled within the project. This paper therefore provides an overview of how to structure a value assessment and the attributes that need to be considered. It aims to provide guidance to support interested stakeholders who wish to make an assessment of the advantages and disadvantages of different thermal treatment technologies.

2. Structuring a value assessment exercise

When conducting a value assessment, it is important to follow a structured approach that focuses the assessment before it is undertaken. As outlined in Figure 1, the first stage in this process is to choose the specific waste(s) and technologies that will be assessed. Once the candidate waste(s) and technologies have been chosen, the assessment scope and approach can be clearly defined. This firstly involves deciding on the attributes that are to be considered and the lifecycle stages that are assessed. A decision should also be made at this stage on whether the assessment will be conducted as a narrative (summarising available information and assessment outcomes) or whether scoring will be applied to allow the relative value of the different technologies to be quantified. Should the assessor wish to apply some quantitative scoring, this should be carefully calibrated so that no technology is unduly favoured.

Once the scope of the assessment has been defined and the approach has been chosen, information should be gathered about both the nature of the waste and the properties of the technology. The amount of information required, depends on the level of detail needed for the assessment and the number of simplifying assumptions that can be tolerated.

It is easiest to perform the assessment on a comparative basis, where the chosen waste – technology combinations are compared against a non-thermal waste management baseline, such as grout encapsulation. Such a comparative evaluation allows the advantages and challenges of each waste – technology combination to be clearly identified across the full waste lifecycle (i.e., from retrieval of raw waste through pre-treatment, conditioning, packaging, storage and disposal). A comparative evaluation also allows exclusion of management steps for which there is no differentiation between the thermal and the non-thermal baseline, thus simplifying the value assessment process. Therefore, sufficient information on the baseline technology should also be available to the assessor to allow a full and balanced comparison to be made.

IOP Conf. Series: Materials Science and Engineering **818** (2020) 012003 doi:10.1088/1757-899X/818/1/012003

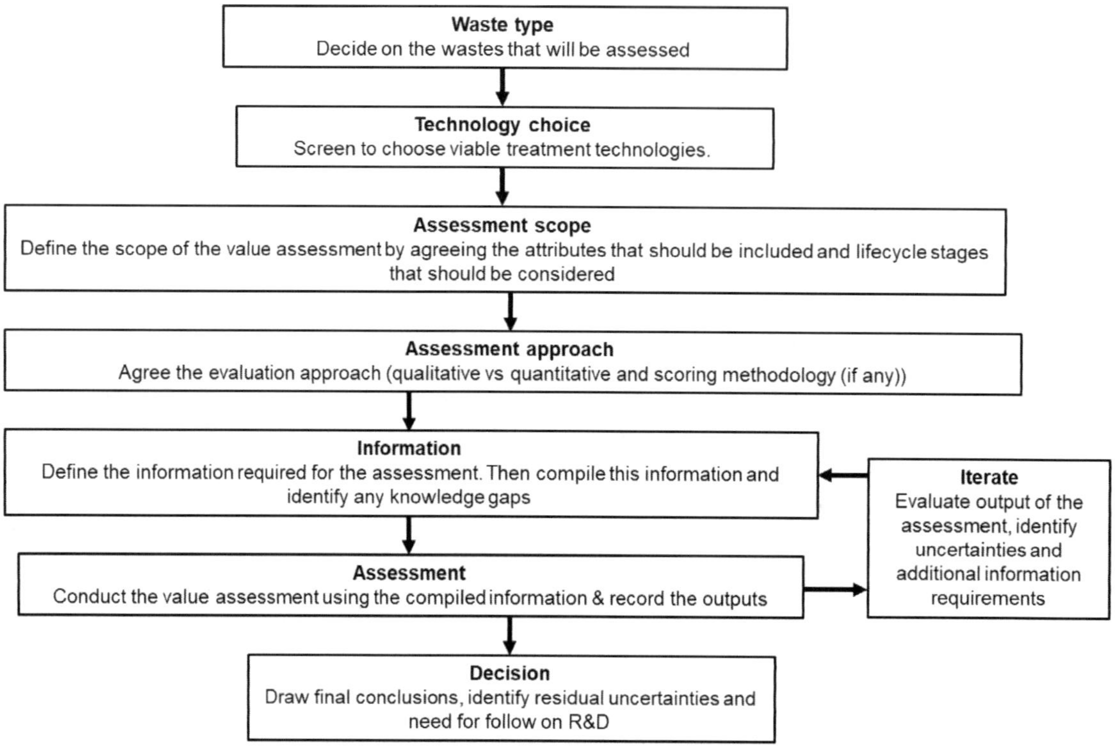

Figure 1. Flow chart summary of the structured value assessment process

3. The value assessment attributes

The THERAMIN value assessment is based around a series of attributes that are sub-divided into data categories. As part of defining the scope of the assessment, the assessor should decide on which of these attributes to include. The attributes chosen in the THERAMIN project are outlined in Table 1.

A decision should also be made on the lifecycle stages to which the attributes apply. The lifecycle stages in this context aim to capture differences between attributes at different times during the construction and operation of the facility and management of the waste. For example, different operational safety concerns may arise during construction compared to operations. Some attributes, such as technological readiness and timescales cover the whole lifecycle and may not need to be fully differentiated. The lifecycle stages of relevance to the value assessment are:

- Facility design and construction
- Waste pre-treatment (such as size reduction)
- Treatment operations
- Waste post-treatment processes (such as conditioning)
- Storage and disposal of the treated waste product
- Decommissioning of the facility at end of life

Table 1. Summary of assessment attributes chosen to assess thermal treatment technologies.

Attribute	Data Category
Operational and Transport Safety	Facility construction and decommissioning
	Waste pre-treatment requirements (conventional and radiological safety implications)
	Waste post-treatment requirements (conventional and radiological safety implications)
	Waste operational safety issues (e.g., ease of providing shielding during operation)
	Transport safety issues
Environmental Impact	Material requirements
	Energy requirements
	Secondary waste and gaseous/liquid discharges generated
	Nuisance
Impact on disposability/ long-term safety	Ability to meet waste acceptance criteria
	Disposability of secondary waste
Implementation	Indicative lifetime feed
	Ease of achieving required throughput for process (full-scale facility) (m^3/year)
	Potential to treat a wide range of waste groups (flexibility) including problematic and orphan wastes
	Impact on waste management strategy
Timescale	Design, construction and active commissioning timescale
	Lifetime operating timescale
	Decommissioning timescale
Technical Readiness	Maturity of the technology
Strategic Cost Impact	Costs of construction, operation and decommissioning
	Impact on disposal costs (total packaged waste volume and required storage and disposal capacity)

The following sub-sections give further detail on the different attributes and the sorts of considerations that should be made when assessing thermal treatment facilities.

3.1. Operational and transport safety

When conducting any waste treatment activity, it is important to consider the safety of the workers involved in the process and the hazards to which they might be exposed. This attribute should therefore assess the intrinsic hazards associated with each technology during treatment operations. This includes both conventional hazards, such as burns, as well as the radiological doses to which the workers are exposed. Although it is assumed mitigating actions will be taken to manage any hazards and ensure worker safety, these will still be present and should be accounted for during any assessment.

Different safety concerns are likely to be present at different lifecycle stages, therefore separate consideration should be given to the hazards associated with the construction and decommissioning of the facility infrastructure in addition to those hazards that are present during operations (including pre- and post-treatment, if required). Additional consideration should be given to safety issues associated with transport of the waste, both to and from the facility, for example the risk of road traffic accidents involving transport lorries.

3.2. Environmental impact

This attribute group considers both the local and wider environmental impact of operating the chosen waste treatment technology. This includes both direct impacts of the facility, such as discharges, as well as indirect impacts associated with the use of materials and energy. Consideration should be given to all the (chosen) lifecycle stages including construction (and decommissioning) of the facility infrastructure, as well as operations.

It is assumed that the environmental impacts of any authorised facility will fall within regulatory limits, however their impact should still be considered, as it may act as a differentiator between technologies.

3.3. Disposability

This attribute assesses the final disposability of the treated waste product. This can be assessed either against country (and facility) specific waste acceptance criteria, where available, or done generically. A series of generic waste acceptance criteria have been developed in work package 4 [3,4], and these are made available for the assessor to choose from (should no facility specific criteria be available). It is left up to the assessor's discretion to decide which of these are included and which are deemed out of scope.

- Dimensions / mass of packages
- Provisions for handling, transport and emplacement
- Package integrity and required lifetime
- Activity content
- Radionuclide inventory
- Dose rate limits
- Surface contamination
- Nuclear criticality
- Thermal output
- Gas generation
- Chemical content
- Chemical durability
- Voids
- Waste package stacking
- Waste package impact performance
- Waste package fire performance
- Identification / labelling
- Quality assurance / quality control requirements
- Data management
- Secondary waste

Should the assessor wish, this disposability criteria could be replaced or amended to assess the storage of the waste product in addition to or in place of final disposal. This may be the preferred option if a final disposal facility is not available and interim storage of the product is required.

3.4. Implementation

This attribute focuses on the strategic implications of implementing the technology, including considering the ability of the treatment technology to meet the required need of the waste management programme. Particular relevance should be paid to whether the technology could provide sufficient lifetime capacity, and at an acceptable throughput rate to process the waste stream being assessed in a timely manner. It is assumed that if any technology is implemented it will be designed to fulfil the requirements of the end user. Therefore, this attribute principally applies to the design and construction stage of the lifecycle. Specifically, the assessor should consider how easily the technology can be scaled up or down to achieve the desired throughput. In addition to those properties of a technology that can be designed to meet requirements, each technology will also have inherent limitations, such as its ability to treat a range of wastes. It would be up to the assessor to decide whether they wish to use a flexible technology that can treat a range of waste types or a bespoke piece of equipment to target a specific waste stream. Finally, the impact on the waste management strategy should consider the wider implications of using the technology for the entire lifecycle of the waste including changes to decommissioning timescales and disposal practices.

3.5. Timescale

The timescale attribute assesses the time required to implement the technology and how long it would operate for. Again, consideration should be given to the whole lifecycle. This includes time required to design and construct the facility (is it off-the-shelf or a one-off), the lifetime of the facility and how much of that lifetime it would be operational (vs offline for maintenance), and then how long it would take to decommission and demolish. This attribute could also be used as a basis to assess whether the chosen technology could fit into wider strategic timelines, across the whole waste management programme (and lifecycle).

3.6. Technical readiness

The technical readiness of a technique is a measure of how readily it could be implemented at the full industrial scale. The technical readiness of a specific thermal technology can be broadly judged at the design stage as fitting into one of three broad groups:

- a prototype still undergoing active research
- deployed at the pilot scale
- deployed at the full industrial scale

If desired, greater granularity can be provided by using a scale of technology readiness levels, which generally range from TRL 1 (established in principle) to TRL 9 (operating industrially) [5].

The technical readiness of a technique will impact both the timescale and cost of implementing it. For example, a less mature technology could require additional R&D or testing to allow it to treat the chosen waste stream. This would necessitate an investment of time and money prior to implementation. Conversely, a fully proven industrial technique could be implemented immediately.

3.7. Cost impact

This category considers the overall cost of constructing and operating the facility, over its full lifecycle. This includes the initial capital costs of infrastructure and construction, and well as operational and decommissioning costs. These includes material and energy costs, as well as the labour costs (impacted by the number of staff required). In addition to these direct cost impacts, the indirect cost impact of treating the waste on the wider waste management programme could also be assessed. Specifically, the change in volume or activity of the product may impact the required storage and disposal capacity, and potentially the nature (and cost) of the stores or disposal facilities.

4. Summary and conclusions

The value assessment methodology presented in this paper was developed from the NDA Value framework [1] though work undertaken in WP2 of the THERAMIN project. It incorporates the output of the THERAMIN value assessment workshop that considered the application of different thermal treatment technologies to various types of radioactive wastes. The assessment attributes have been developed so that they address all the lifecycle stages and the priorities of the different partners (technology owners, waste producers and waste management organisations).

The methodology summarised in this paper is provided as a generic starting point that can be tailored to the needs of the assessor. It is recommended that before conducting any assessment, the required objectives should be clearly defined and the assessment approach tailored to achieve these. For example, a decision should be made on whether the output should be a qualitative narrative (simply listing the advantage and challenges associated with each technology) or include a quantitative scoring. Additionally, the level of detail required in the output should also be decided, so that sufficient information can be gathered to support the analysis. The level of information required is likely to depend on the stage in the decision process that the assessment is being done. For a scoping assessment (on a wide range of options) the relatively small amount of information available in the public domain may be sufficient to allow a high-level judgement to be made. However, at later stages in the process where more robust and fully supported arguments are required, a greater level of detail would be necessary.

Overall this paper should act as a starting point for those new to the value assessment process in the context of thermal treatment. It is designed to encourage structured decision making about the implications of utilising thermal treatment technologies for radioactive waste management. Extended detail on the process can be found in THERAMIN deliverable D2.5 [6], on which this paper is based.

Acknowledgements
This project has received funding from the Euratom research and training programme 2014-2018 under grant agreement No 755480. This paper reflects the authors' views and the European Commission is not responsible for any use that may be made of it.

References

[1] Nuclear Decommissioning Authority 2016 The NDA Value Framework, Version 1.2
[2] Andra et al. 2018 Waste Acceptance Criteria and requirements in terms of characterisation, EC THERAMIN D4.1
[3] Wareing A, Abrahamsen-Mills L, Fowler L, Grave M, Jarvis R, Metcalfe M, Norris S and Banford AW 2017 Development of integrated waste management options for irradiated graphite. *Nuclear Engineering and Technology*, 49(5), pp.1010-1018
[4] Harvey E J, Galson D A, Catherin S, Romero M-A, Fournier M, Fuller A J and Wickham S M 2020 Development of generic criteria for evaluating the disposability of thermally treated wastes *IOP Conf. Ser.: Mater. Sci. Eng* same issue
[5] Nuclear Decommissioning Authority 2014 Guide to Technology Readinesss Levels for the NDA Estate and its Supply Chain, 13594
[6] Galson Sciences Ltd 2020 Value Assessment, EC THERAMIN D2.5, to be published

IOP Conf. Series: Materials Science and Engineering **818** (2020) 012004 doi:10.1088/1757-899X/818/1/012004

Active demonstration of the thermal treatment of surrogate sludge and surrogate drums using the GeoMelt™ In Container Vitrification (ICV) melter installed in NNL Central Laboratory

S Clarke[1], C Scales[1], N Patel[2], J Roe[1] and A W Banford[2]

[1] National Nuclear Laboratory, Central Laboratory, Sellafield Works, Seascale, CA201PG, United Kingdom
[2] National Nuclear Laboratory, 5th Floor Chadwick House, Birchwood Park, Birchwood, Warrington, WA3 6AE, United Kingdom

sean.clarke@uknnl.com

Abstract. Under the auspices of the THERAMIN project, the GeoMelt™ in-container vitrification (ICV) melter system installed in the National Nuclear Laboratory (NNL) Central Laboratory at Sellafield, UK, has been used to demonstrate the immobilisation of two surrogate feeds. The streams selected for demonstration were a cementitious stream representative of legacy drums or failing cemented wastes packages and a wet waste stream made up of a surrogate for Magnox storage pond sludge and clinoptilolite, an ion exchange material, as a way of demonstrating co-immobilisation. The distribution of waste species throughout the vitrified block were assessed to determine the efficacy of the mixing during the thermal treatment process. Analysis of samples from the off-gas system was carried out to establish an activity balance across the process. This paper will outline the process set up, provide operational data from the melt concluding with an assessment of the applicability of the trials results to thermal treatment of these and similar waste streams.

1. Introduction

In the UK, High Level Waste (HLW) and Intermediate Level Waste (ILW) have been successfully immobilised over many decades. HLW has been treated using vitrification and some operational ILW streams by immobilisation in a cement matrix.

Cementitious encapsulation was selected as the baseline treatment technology for many ILW was selected as the baseline treatment technology as it is simple, readily available and is a low-cost material compatible with wet wastes. However, ILW destined for cementation does not generally undergo pre-treatment. In the absence of this stage, organic plutonium contaminated material (PCM) and/or reactive metals could retain chemical reactivity within cementitious products; thus, additional considerations are required for interim storage and final disposal.

Thermal treatment offers an alternative waste treatment process. The application of heat in a thermal process can pacify the waste metals and provide significant volume reduction, this leads to a product that is potentially more stable and more cost effective for storage and disposal.

In the UK, NNL plays a key role in the technical integration and maturation of technologies through the Technology Readiness Levels (TRLs), towards subsequent deployment by end users. The

GeoMeltTM system at the NNL Central Laboratory [1] at Sellafield was commissioned to treat waste and prove the concept that thermal treatment technologies could be chosen as a viable alternative to grout encapsulation for the immobilisation of some ILW streams. Progress made in recent years in this area has included advances with respect to not only the operation of such a system in an active environment, but also advances with respect to the safety case approach required to do so. NNL has published a position paper on thermal treatment of ILW which provides further background [2]. Learning from this experience within NNL formed a key element of NNL's leadership of the thermal treatment demonstration work package (WP3) of the THERAMIN project.

2. In Can Vitrification and GeoMelt™ at NNL Central Laboratory

GeoMelt™ incorporates a range of patented and proprietary vitrification processing technologies that are configured in a variety of ways depending on the waste to be treated and the disposal route (in-situ or containerised). GeoMelt™ transforms hazardous chemical and radioactive wastes into an inert vitreous monolith. Hazardous organics are destroyed by pyrolysis, and heavy metals and radionuclides are immobilized within the glass matrix.

The basis of heating within the GeoMelt™ thermal system is joule-heating via four electrodes. The joule-heating principle operates by internal resistance heating of conductive material as the electrical current passes through. The GeoMelt™ process is initiated at a relatively high resistance requiring a higher voltage potential and lower current. As the melt progresses and resistance decreases, lower voltage taps allow increased current to the melt.

The GeoMelt™ configuration is illustrated in Figure 1. The melter (1) is powered by a 150kW transformer. As the melt progresses, additional material can be fed using the feed while melt system (FWM - 2,3) to take advantage of the space left by volume reduction of the initial waste and reduce voidage within the CRB. The entire system is held under ventilation depression during melting using a blower with a contingency back up (12,13), air is drawn through a filtered inlet line (4) The first stage of the off-gas treatment system is a sintered metal filter (SMF - 5). This is back pulsed during operation to prevent blockage and the associated increase in differential pressure.

Figure 1. GeoMelt™ ICV System.

IOP Conf. Series: Materials Science and Engineering **818** (2020) 012004 doi:10.1088/1757-899X/818/1/012004

Following the SMF, the off-gas stream passes through a cooled caustic scrubbing system (6,7,8,11). The off-gas stream requires heating following scrubbing to avoid moisture dropping out and corroding the pipework, it is heated in line (9). The final stage of filtration is provided by two in line HEPA filters (10) before connecting to the building ventilation network (14). The NNL Central Laboratory GeoMelt™ ICV installation is shown in Figures 2 and 3.

Figure 2. GeoMelt™ ICV at NNL Central Laboratory.

Figure 3. GeoMelt™ ICV installed at NNL Central Laboratory.

3. THERAMIN Demonstration Trials

3.1 Trial Feed Composition

Two specific waste streams were selected for the THERAMIN experimental programme based on the THERAMIN WP2 assessment. Both the wastes tested were identified as being of interest both in the UK and internationally and are referred to subsequently as TH-01 and TH-02.

- TH-01 - A cementitious stream representative of drums or failing cement wastes packages
- TH-02 - A wet waste stream made up of a naturally occurring zeolite (clinoptilolite), sand, Magnox storage pond sludge and miscellaneous contaminants known to arise in a range of UK feed streams.

These are detailed in Table 1.

IOP Conf. Series: Materials Science and Engineering **818** (2020) 012004 doi:10.1088/1757-899X/818/1/012004

Table 1. Feed Compositions of GeoMelt™ trials TH-01 and TH-02.

Trial TH-01	Mass (kg) unless stated otherwise	**Trial TH-02**	Mass (kg) unless stated otherwise
Mild steel tins	3.15	CMgS Sludge (45 % v/v solids)	50.25
Cement	13.35	Clinoptilolite	83.8
Aluminium	2.45	Aluminium	1
Plastic (PVC)	3.05	Iron	1
		PAG Oil	1
Glass Formers			
Quarry soil, Boric acid, Soda carb + VTR frit blend	176.6	Silica	8.4
		Sodium Carbonate	16.7
		Boric Acid	8.4
Tracers & Starter Path			
Cs-137 (Active)	25 MBq	Cs-137 (Active)	25 MBq
$SrCO_3$ (Non-active)	(8.42 g)	Sr-85 (Active)	16.1 MBq
CeO_2 (Non-active)	(6.14 g)	EuO_2 (Non-active)	5.8 g
		CeO_2 (added with feed via FWM system)	6.14g
Glass frit (with starter path)	20	Glass frit (with starter path)	20
PRE-STAGE TOTAL	**218.6**		**190.55**
Feed While Melting			
Glass frit	50	Glass frit	12
Soil	10.65		
Total Material	**279.25**		**202.55**

3.2 Operational Observations

3.2.1 TH-01. Melt TH-01 was carried out over a 15 h period. In total 279.25 kg of material was processed at an equivalent rate of 18.6 kg/h. The melt video can be seen here [8].

At several points throughout the melt significant gas evolution was observed at the surface, this was consistent with the decomposition of grout and PVC. This was observed in phases consistent with the melting of each successive layer of tins/cans. At one point, tins were observed to be floating on the surface of the melt, which could have potentially caused a problem of electrical shorting across the electrodes. However, addition of cold feeds dispersed the tins which mitigated this risk. The SMF pressure differential peaked at ~15 inWG with the blower operation peaking at ~80% of its maximum output. Peak power was 34 kW/phase, this was slightly higher than predicted but the higher power was required to reach the target temperature of 1350oC. The melter hood temperature was maintained below its operational control point high of 550oC and the electrical parameters for both phases were well balanced throughout the melt. Operator observation, via the infrared (IR) camera, indicates that adequate convection and movement occurred suggesting good mixing within the melt pool.

3.2.2 TH-02. An initial attempt to melt the feed material was unsuccessful due to the rapid volume reduction of the starter path material which led to a drop of melt surface material away from the electrode tips before they could be lowered. This led to the loss of the conductive path and melt progression. Once cooled the CRB was prepared for re-melt through the introduction of a fresh starter path [3].

The TH-02 re-melt was carried out over 15 h, in total 202.55 kg of material was processed at an equivalent rate of 13.5 kg/h. The melt video can be seen in [9].

At 10.5 h into the melt one of the joule heating phases went into fault. It was not immediately obvious at the time what had happened. The operational decision at this point was to continue processing using the remaining operational phase. Only post-melt was the cause of the fault determined. On retrieval of the product it could be seen that the lowest segment of one of the electrodes had fallen away from the segment above it, breaking the electrical circuit. Due to the loss of one heating phase, no further feeding was undertaken, this was to minimise the challenge on the melter. The power on the remaining phase was increased to 34 kW to compensate for the absence of the second phase, the melt reached its target temperature of 1260oC. The most probable cause of the detachment of the electrode segment is that that the segments were not coupled together robustly during the electrode installation phase of the melt cycle.

4. Results

4.1 Product Appearance

Figures 4 and 5 are pre and post melt images from experiment TH-01, Figures 6 and 7 for experiment TH-02.

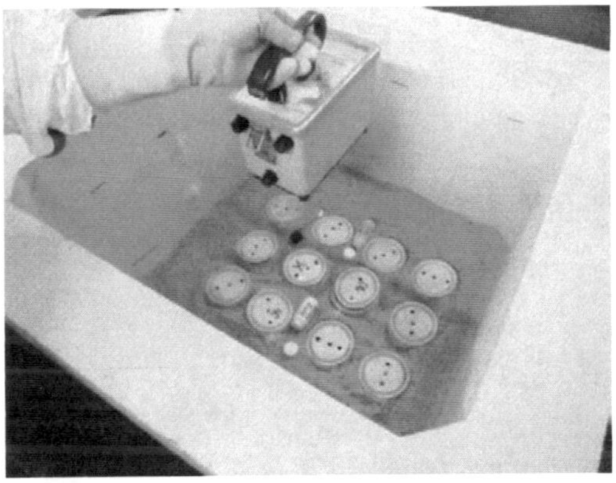

Figure 4. TH-01 during loading.

Figure 6. TH-02 starter path.

Figure 5. TH-01 vitrified product.

Figure 7. TH-02 vitrified product.

IOP Conf. Series: Materials Science and Engineering **818** (2020) 012004 doi:10.1088/1757-899X/818/1/012004

4.2 Quantities of Product and Secondary Waste

4.2.1 TH-01. Melt TH-01 was cooled in situ and then the ICV and product was removed from plant. On removal of the product, the off-gas pipe connecting to the melter hood was inspected for obstructions, but none were found. Solids were removed from the SMF filter, in total 1.085 kg of material was recovered. In addition, a sample of the scrub tank liquor was taken. A 200 ml sample was taken from a bulk tank volume of 2000 l. SMF solids and scrub tank liquor were submitted for radiochemical and chemical analysis.

The bulk product appeared to have a relatively flat glassy surface, looked broadly homogenous and visually appeared to be a satisfactory product. The electrodes had thinned significantly; this is normal and is observed in virtually all melts carried out with this equipment. The mass of the glass monolith within the CRB was measured to be 236 kg.

The reduction of volume due to the thermal process and the ability to feed additional material during the melt can result in a net overall waste volume reduction for disposal. The potential life cycle benefits of reducing waste volumes for geological disposal during reactor decommissioning was identified by Wallbridge et al [4].

4.2.2 TH-02. As in TH-01, melt TH-02 was cooled in situ and the ICV and product was removed from plant. Solids were removed from the SMF filter, in total 0.16 kg of material was recovered. This amount of material can be considered normal for this process. A sample of the scrub tank liquor was taken. A 200 ml sample was taken from a bulk tank volume of 2000 l. SMF solids and scrub tank liquor were submitted for radiochemical and chemical analysis.

When the product monolith was presented for sampling, the bulk product appeared to have a glassy surface and appeared to be a satisfactory product. As in TH-01, the electrodes had thinned significantly. The mass of the glass monolith within the CRB was measured to be 197 kg. The relatively low product mass resulted in a product container which approximately half full, this is not typical but was due to the inability to feed more than 12 kg during the melt cycle due to the failure of one of the heating phases as described earlier.

4.3 Radioactivity Balance

Glass product and off gas samples taken from melts TH-01 and TH-02 were prepared for Gamma spectroscopy analysis. For each glass sample point a 10g aliquot was finely ground and gamma scanned to determine the Cs-137 and Sr-85 activity concentrations. Similarly, a 10g aliquot of the filtered solids collected in the SMF and 10ml aliquot of the scrub tank liquor were also gamma scanned to determine the Cs-137 and Sr-85 activity concentrations. Table 2 below outlines the results of this analysis. Whilst not available for the TH-01 and TH-02, muon tomography has also been used to provide additional analytical data on GeoMeltTM products from other trials as detailed by Simpson et al [5].

Table 2. Radioactivity analysis across GeoMelt™ System following TH-01 and TH-02.

Melt, Radioisotope	Activity in glass monolith (%)	Activity in SMF solids (%)	Activity in Scrub tank liquor (%)	Difference (%)
TH-01 Cs-137	76.0	1.9	1.2	20.9
TH-02 Cs-137	76.9	0.01	1.2	21.9
TH-02 Sr-85	76.9	>0.01	0.38	22.8

Activity measurements in both TH01 and TH02 trials have yielded Cs retention levels consistent with expectations for a one-off unoptimized melt. It is considered that operational changes along with glass composition modification could significantly increase Cs retention. Sr-85 losses appear high and may be due to errors in the gamma spectroscopy measurements or could indicate entrainment of solids rather than volatility.

Most of the activity in the off-gas system is confined to the SMF, which allows for capture and reprocessing through subsequent melts. Mass and activity analysis across the glass blocks show consistent results indicating that the product is homogeneous, and that waste feeds and additives have been well reacted during the melt process.

The difference in activity identified in Table 2, is probably due to activity retained with the rig system possibly due to a combination of plate out and particulates. This will be investigated in future trials.

5. Conclusions

The active trials carried out by NNL using the GeoMelt^{TM} system installed at the NNL Central Laboratory in Sellafield have successfully demonstrated that thermal treatment can in principle be used to treat surrogate ILW wasteforms. The trials also demonstrate the ability to successfully immobilise both heterogeneous and homogeneous wastes through thermal treatment.

The findings and learning from this proof of concept programme are promising, however it must be stressed that the thermal treatment of wasteforms such as this could be further optimised. The key area of focus would be the optimisation would be changes to the process required to achieve a much higher activity retention % in the final product. If thermal treatment was chosen to treat a given wasteform an in depth and sustained programme of trials would be required to fully understand the feed, optimise the process and maximise the waste loading and volume reduction. Previous experience within NNL in support of the thermal treatment of highly active liquor from spent fuel reprocessing to produce vitrified HLW for Sellafield Ltd. is a useful relevant example of this recommended approach [6].

THERAMIN trials TH-01 and TH-02 using GeoMelt^{TM} has provided an opportunity to demonstrate progression up the Technology Readiness Level (TRL) scale for both feeds from 3 – 6 [7].

Acknowledgments

This project has received funding from the Euratom research and training programme 2014-2018 under grant agreement No 755480. The authors would also like to acknowledge the funding received under the NNL Thermal Treatment Core Science programme. The paper reflects the author's view and the European Commission is not responsible for any use that may be made of it.

References

[1] Campbell B, Dysland E J, Finucane K G, Hatt S, Moulin-Ramsden M, Woosley S L, Scales C R and Roe J I 2017 Commissioning of a GeoMeltTM vitrification plant at the Central Laboratory on the Sellafield site, WM2017 Conference, Phoenix, Arizona, USA

[2] Angus M, Gribble N and Scales C R, 2015. Thermal processes for immobilising intermediate level wastes. https://www.nnl.co.uk/innovation-science-and-technology/121-2/position-papers/#thermal-processes-for-immobilising-intermediate-level-wastes

[3] Roe J I, Burns K, Patel N, Scales C R, Clarke S, 2019. Test report on demonstration of GeoMeltTM technology for THERAMIN project, D3.5 NNL.

[4] Wallbridge S, Banford A and Azapagic A, 2013. Life cycle environmental impacts of decommissioning Magnox nuclear power plants in the UK. The International Journal of Life Cycle Assessment, 18(5), p990-1008

[5] Simpson A, Clarkson A, Gardner S, Jebali R A I, Kaiser R, Mahon D, Roe J, Ryan M, Shearer C and Yang G, 2020. Muon tomography for the analysis of in-container vitrified products. Applied Radiation and Isotopes 157

[6] Short R, Gribble N, Turner E, Riley A D, 2010. Using the vitrification test rig for process improvements on the waste vitrification plants. *Advances in Science and Technology* **73**
[7] Guide to technology readiness levels for the NDA estate and its supply chain, November 2014,https://assets.publishing.service.gov.uk/government/uploads/system/uploads/attachment_data/file/457514/Guide-to-Technology-Readiness-Levels-for-the-NDA-Estate-and-its-Supply-Chain.pdf
[8] THERAMIN GeoMelt TH-01, https://youtu.be/QNw4Wo0QXPM
[9] THERAMIN GeoMelt TH-02, https://youtu.be/Y5caa8CW3X0

IOP Conf. Series: Materials Science and Engineering **818** (2020) 012005 doi:10.1088/1757-899X/818/1/012005

In-Can vitrification of ash

M Fournier, N Massoni and J F Hollebecque

CEA, DEN, DE2D, Univ. Montpellier, Marcoule, F-30207 Bagnols-sur-Cèze

maxime.fournier@cea.fr

Abstract. The In-Can Melter is a metallic crucible heated in a refractory furnace using electrical resistors allowing in-container vitrification. The In-Can Melter trial aims to demonstrate the feasibility of the confinement in a glassy matrix of ash coming from existing incineration processes. The ash was pelletised to allow its introduction into the can without dust emissions and then incorporated in a 50 wt.% waste loading confinement matrix. The full-scale trial was preceded by laboratory- and bench-scale tests. The microstructure and chemical durability of the wasteform were characterised.

1. Introduction

The Thermal treatment for radioactive waste minimisation and hazard reduction (THERAMIN) project is a European Commission programme of work jointly funded by the Horizon 2020 Euratom research and innovation program and various European nuclear waste management organisations (WMOs). The THERAMIN project is being conducted between June 2017 and May 2020. Twelve European WMOs and research and consultancy institutions from seven European countries are participating in THERAMIN. The overall objective of THERAMIN is to provide improved, safe long-term storage and disposal of intermediate-level waste and low-level waste suitable for thermal processing. The work programme provides a vehicle for coordinated European Union-wide research and technology demonstrations designed to provide improved understanding and optimisation of the application of thermal treatment in radioactive waste management programmes across Europe, and moves technologies higher up the Technology Readiness Level (TRL) scale.

In the framework of the THERAMIN project, the French Alternative Energies and Atomic Energy Commission (CEA) carried out studies on the treatment and conditioning of ash resulting from the incineration of technological surrogate waste. The trial described in this article is based on the CEA In-Can Melter (ICM) consisting of a metallic crucible melter heated in a refractory furnace. Prior to the full-scale trial, laboratory- and bench-scale tests were conducted to select optimised operating conditions. Wasteforms produced were characterised after the trials.

2. Description of the CEA's In-Can Melter

The ICM is a metallic crucible heated in a refractory furnace using electrical resistors (Figure 1). The can, with an approximate volume of 50 L, is renewed after each filling. The process can support either liquid or solid waste feeds. With the current off-gas treatment system (not including a post-combustion chamber), it can only tolerate small amounts of organics. It can also accept a small fraction of metal in the waste. The design ensures that the process can operate remotely for high-activity waste. The design can also be adapted for dealing with plutonium-containing material in gloveboxes. The end product can be glass, glass ceramic, or simply a high-density waste product.

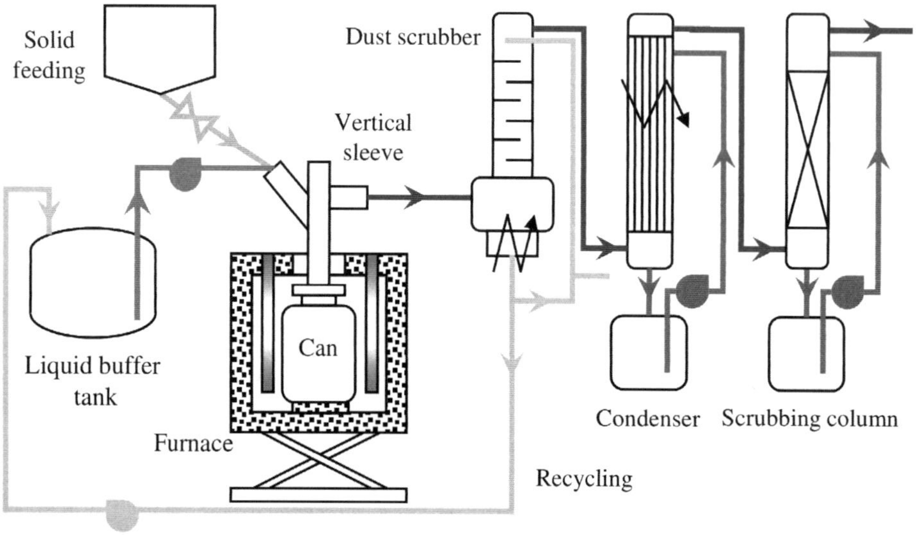

Figure 1. Diagram of the CEA's In-Can Melter.

ICM has been developed for CEA needs since 2005 for military waste vitrification. The technology is at TRL 7 and the ICM has undergone inactive commissioning in France with inactive simulants of alpha effluent. The technology is currently being qualified for solid waste.

3. Ash production

The waste selected for this vitrification trial is the ash from multiple incineration tests of surrogate technological waste (polyvinyl chloride, latex, neoprene, polyethylene, cotton, etc.) produced by the CEA's IRIS process (French acronym for Research facility for solid waste incineration). IRIS is a research facility for the incineration of solids developed to treat organic waste from gloveboxes in the nuclear industry, contaminated with alpha-bearing actinides and containing high quantities of chlorine (Figure 2).

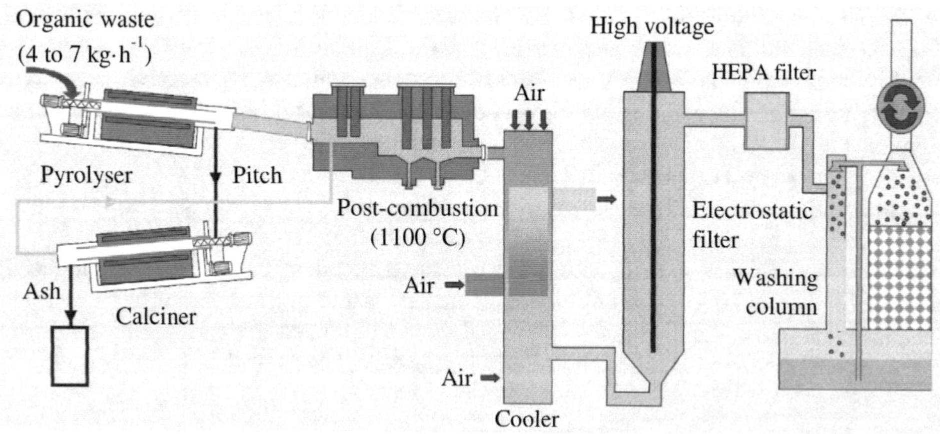

Figure 2. Diagram of the CEA's IRIS process for the treatment of organic waste.

The robustness and efficiency of this process is based on a decoupling of (*i*) the step of elimination of corrosive materials such as chlorine and (*ii*) the step of combustion of the organic waste. Figure 2 shows that the organic waste first goes into a pyrolysis step at a temperature of 500 °C to remove the most corrosive gaseous compounds and then into a calciner (900 °C) fed with oxygen to complete the combustion while concentrating the contamination in the mineral ash. The relatively long residence time in furnaces with low gas flow rates makes it possible to produce carbon-free ash, concentrating almost all of the initial activity. The off-gas treatment system, consisting of a post-combustion chamber followed by electrostatic filtration, ensures excellent purification.

4. Pre-treatment of ash and definition of their vitrification conditions

In order to obtain a sufficient amount of ash to carry out the various tests described thereafter, 230 batches, each containing approximately 220 g of ash from various IRIS trials, were homogenised using a Morton Mixers ribbon blender (Figure 3.a-b). This operation resulted in a decrease in the size of the ash agglomerates, with an ash apparent density of 0.6 $g{\cdot}cm^{-3}$, compared to 0.2 $g{\cdot}cm^{-3}$ before homogenisation, and a true density of 1.9 $g{\cdot}cm^{-3}$ measured by helium pycnometry. The composition of the ash thus mixed (Table 1) was analysed by X-ray fluorescence spectroscopy (XRF).

Ash is characterised by a high volatility that can lead to clogging of the feeding and off-gas treatment system pipes and large dust carry-over. In order to limit this volatility, a temporary densification by pelletisation was carried out using a Frogerais rotative press. The ash was pressed in the presence of a binder to ensure the mechanical cohesion of the pellets, but without glass frit, being too abrasive for the press. Two types of binder were tested: organic (sorbitol, $C_6H_{14}O_6$, or fructose, $C_6H_{12}O_6$) and mineral (bentonite, mainly containing montmorillonite $(Na,Ca)_{0.33}(Al,Mg)_2Si_4O_{10}(OH)_2(H_2O)_n$). The addition of 10 wt.% of fructose or bentonite to the ash gave similar and satisfactory results. Bentonite was preferred thereafter to avoid possible CO_x gas emissions during combustion. Approximately 60,000 pellets (9 kg of ash), reaching a 72% densification, were produced (Figure 3.c).

Figure 3. Ash in the mixer (a) before and (b) after homogenisation. (c) Ash pellets.

In order to determine the vitrification conditions of the ash, laboratory-scale tests, using small quantities of materials (≈10 g), were carried out with different amounts of ash and glass frit at 1100 °C for 2 h. In the approach proposed in the THERAMIN project, no vitreous matrix formulation study was carried out and FN0C77 glass frit (Table 1), simple and available, was used. At the end of the tests, the crucibles were cut and the materials obtained were observed with a binocular magnifier and evaluated according to two criteria: (*i*) obtention of a macroscopically homogeneous material and (*ii*) the absence of foaming during melting. The best results were obtained for a 50:50 mixture between ash and glass frit. Beyond this ratio, the reactivity between ash and glass frit is less efficient and the resulting material becomes less homogeneous.

5. Bench-scale test and characterisation of the wasteform

5.1. Test conditions

Following the laboratory-scale tests, the quantity of treated ash was increased during a bench-scale test. This test was carried out in an Inconel™ 601 crucible having an external diameter of 100 mm and a thickness of 5 mm, filled with 400 g of unpelletised ash and 400 g of FN0C77 glass frit (Figure 4.a).

The tapped density of this mixture was estimated at approximately 1 g·cm^{-3}. The mixture was heated at 300 °C·h^{-1} and maintained at 1100 °C for 8 h.

At the end of the test (Figure 4.b), the crucible was cut and the resistance of Inconel™ 601 to corrosion was evaluated. No significant corrosion is seen at the interface between Inconel™ and the wasteform. Above the volume occupied by the wasteform, gaseous species superficially corroded the Inconel™ (Figure 4.c). The crystallised corrosion products are $NiFe_{1.2}Cr_{0.8}O_4$ spinels and nickel oxide. However, no significant reduction in the thickness of the crucible is measured. A mass loss of 2.5 wt.% was measured after the test and a crystallised glass of density 2.63 g·cm^{-3} was obtained (Figure 4.d).

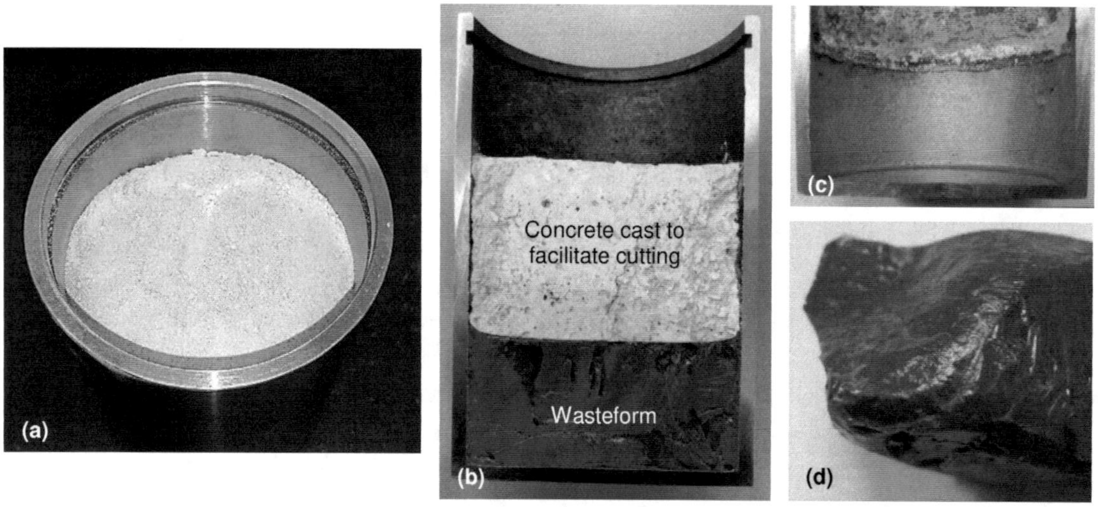

Figure 4. (a) Ash and glass frit mixture introduced into the Inconel™ 601 crucible, (b) crucible cut after the test, (c) inner surface of the crucible after the test and (d) wasteform produced.

5.2. Microstructure and chemical composition of the wasteform

The material is mainly composed of a vitreous matrix consisting of approximately 80 wt.% of Al_2O_3, B_2O_3, CaO, Na_2O, and SiO_2. This matrix includes three types of crystallisation identified by scanning electron microscopy (SEM) and X-ray diffraction (XRD): apatites, zincochromites, and beads composed almost exclusively of bismuth (Figure 5 and Table 1). The stoichiometries of apatites and zincochromites evaluated by energy dispersive X-ray spectroscopy (EDS), respectively $Ca_{4.8}P_{2.9}Si_{0.3}Na_{0.2}Al_{0.1}O_{12.9}$ and $ZnCr_{2.6}Na_{0.5}Mg_{0.2}Al_{0.1}Fe_{0.1}Ni_{0.1}O_{5.8}$ (oxygen assumed for stoichiometry), are consistent with the theoretical stoichiometries of these minerals, respectively $Ca_5(PO_4)_3(OH)$ and $ZnCr_2O_4$. The crystals are homogeneously distributed in the matrix, with the exception of a layer approximately one millimetre thick at the interface with the crucible in which they are more numerous.

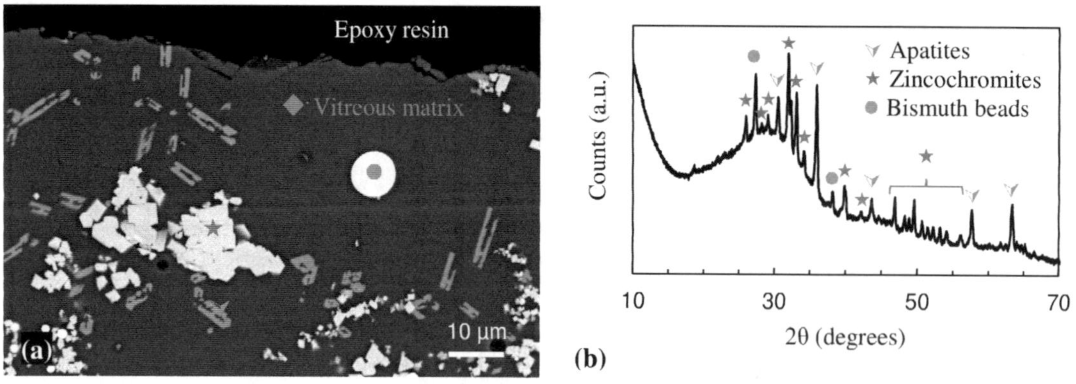

Figure 5. (a) SEM and (b) XRD analysis of the wasteform produced at the bench-scale.

IOP Conf. Series: Materials Science and Engineering **818** (2020) 012005 doi:10.1088/1757-899X/818/1/012005

Table 1. Elemental compositions, expressed in wt.%, of ash and glass frit used for the calculation of the theoretical composition of the wasteform. Analysed compositions of the vitreous and crystalline phases composing the wasteform.

Element	Ash (analysed by XRF)	FNOC77 glass frit (specified)	Wasteform (calculated)	Vitreous matrix ◆ (analysed by EDS)	Apatites ➤ (analysed by EDS)	Zincochromites ★ (analysed by EDS)	Bismuth beads ● (analysed by EDS)
Al	11.90		5.95	9.40	0.06	0.15	
B		10.05	5.02	n.a.[a]	n.a.[a]	n.a.[a]	n.a.[a]
Ba	0.71		0.36				
Bi	4.02		2.01	0.17			9.99
Ca	9.91		4.96	7.10	4.84	0.01	
Cl	1.42		0.71		0.08		
Cr	0.12		0.06			2.58	
Fe	0.50		0.25	0.61		0.05	
K	2.25		1.13	1.56			
Mg	1.59		0.80	3.00		0.24	
Na	0.55	23.74	12.14	13.71	0.16	0.48	
Ni	0.69		0.35			0.05	0.17
P	1.62		0.81	0.62	2.88		
S	0.31		0.16	0.32			
Sb	0.15		0.08				
Si	7.45	16.57	12.01	27.71	0.25	0.02	
Ti	0.45		0.23	1.04		0.03	
Zn	8.13		4.07	5.74		1.00	0.25

[a] n.a.: not analysed because boron is not detected by EDS.

5.3. Chemical durability

The chemical durability of the material was studied according to a protocol adapted from the PCT-B standardised test [1]. After crushing the material, 125–250 µm of powder were placed in contact with pure water with a glass-surface-area-to-solution-volume ratio (S/V) of 10 m^{-1} at 90±2 °C in unstirred perfluoroalkoxy (PFA) reactors. The powder specific surface area of 259 g·cm^{-2} was estimated by adsorption of krypton on the sample surface (Micromeritics ASAP 2020) according to the Brunauer–Emmett–Teller theory [2]. Solution samples, taken at regular intervals, were filtered with a cut-off of 0.45 µm, acidified with ultrapure grade HNO_3 and analysed by inductively coupled plasma-atomic emission spectroscopy (Thermo Scientific iCAP™ 6000 Series). The concentrations were used to calculate normalised mass losses $\mathrm{NL}_i = C_i/(x_i \times S/V)$, where C_i is the concentration of the element i and x_i the mass percentage of i in the glass. The alteration rate, $r = \mathrm{dNL}_i/\mathrm{d}t$, was calculated by linear regression. Boron is known to be an alteration tracer of the vitreous matrix, which means that it is not retained in alteration products while released from the glass.

The evolution of the normalised mass losses of Al, B, Na, and Si is shown in Figure 6.a. Only the elements mainly integrated into the vitreous matrix are considered. The x_i considered are those obtained by EDS analysis for Na, Si, Al, and by calculation for B. Therefore, the dissolution rates of the crystalline phases are not considered, because when a wasteform contains crystalline phases deemed to be durable,

IOP Conf. Series: Materials Science and Engineering **818** (2020) 012005 doi:10.1088/1757-899X/818/1/012005

like apatites or spinels, the wasteform dissolution is controlled by the properties of the vitreous matrix [3]. This leaching test was duplicated, giving similar results: the average values obtained are discussed hereafter.

On the first day of leaching, the alteration rate of the vitreous matrix is ≥ 2.4 g·m^{-2}·d^{-1}. After 3 days, the rate is $\approx$1.5 g·m^{-2}·d^{-1}. The alteration therefore decreases over time and quickly becomes non-stoichiometric, showing the typical behaviour of a borosilicate glass of nuclear interest. Si and Al are retained at 30–40% in the alteration layer after one month. Such retention remains relatively low: the amorphous layer formed on the sample surface therefore remains poor in silicon, which explains why the hydrolysis regime remains predominant. The initial dissolution rate of the vitreous matrix leads to a rapid increase in $pH_{90°C}$ to 8.5, the value at which this parameter stabilises.

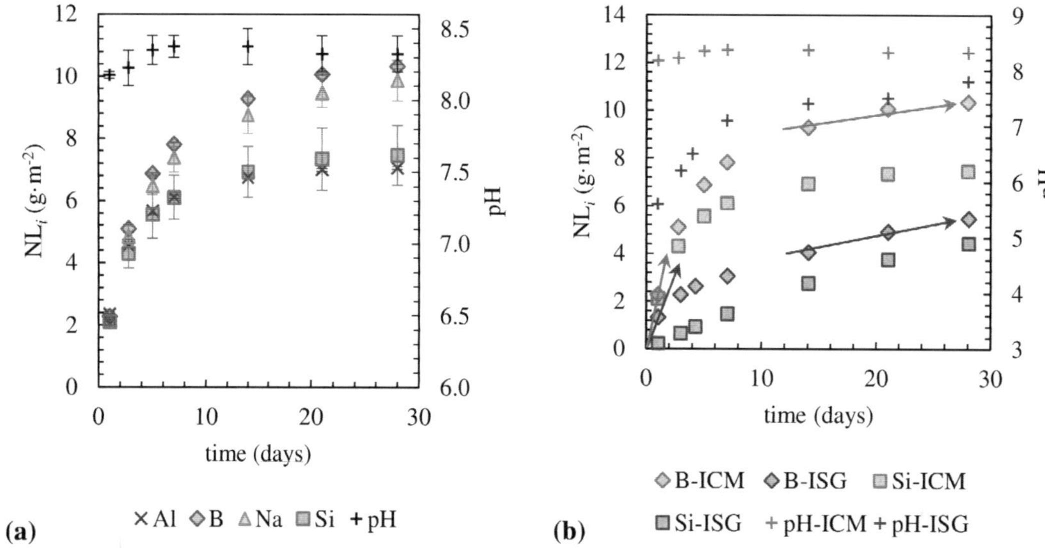

Figure 6. (a) Evolution of NL and pH during wasteform leaching (error bars represent the data dispersion between the duplicates). (b) Compared evolution of NL and pH during the leaching of ISG and In-Can-produced wasteform. The arrows are a visual guide to estimate the drop in alteration rate.

Comparison of previous results with those acquired for the reference International Simple Glass (ISG) [4, 5] altered under the same conditions (Figure 6.b) shows similar and 'classical' trends.

- Because of its composition, the In-Can-produced vitreous matrix has a higher rate of hydrolysis. Indeed, it contains large fractions of Na_2O and B_2O_3, which are unfavourable to the glass durability. Therefore, in the early stages of dissolution, the evolutions of the concentrations in solution are faster for In-Can-produced wasteform. This conclusion is also valid for the pH evolution.
- The growth of the alteration layer is faster for the In-Can-produced wasteform. Thus, its alteration rate drops faster. Between 14 and 28 days, the alteration rate of the In-Can-produced vitreous matrix (0.07 g·m^{-2}·d^{-1}) is almost 1.5 time lower than that of the ISG (0.10 g·m^{-2}·d^{-1}), while the hydrolysis rate of the vitreous phase of the wasteform after 1 day is higher. This result is known for simple glasses [6]: the glasses that are hydrolysed the fastest are also those that have the earliest rate drops.

The data trends tend to show that, in the longer term, an alteration layer will form on the In-Can-produced wasteform, causing a decrease of the alteration rate, as for the ISG.

6. Full-scale trial

A full-scale trial was conducted with the CEA's ICM (Figure 1). The can is made of Inconel™ 601, with an external diameter of 400 mm, a height of 600 mm and a wall thickness of 10 mm.

The trial began with pre-loading of the container with 25 kg of FN0C77 glass frit and 17.8 kg of unpelletised ash. This mixture was heated at 300 °C·h^{-1} up to 1100 °C. Then, 9.2 kg of pelletised ash were fed into the can with a flow of 10 kg·h^{-1}, before the introduction of 1 kg of FN0C77 frit and an additional soaking of 2 h. Thus, the final waste loading was 50 wt.%. The test was continued with 2 h of recycling of the dust recovered in the off-gas treatment system and 2 h of further soaking. The can was then allowed to cool freely. A loss of mass of 2.3 wt.% was measured, consistent with that of the test carried out at bench-scale.

The extent of characterisation of the material produced during this trial is more limited than that of the material produced at bench-scale. However, the SEM and XRD characterisations carried out on a sample taken in the central zone of the can show a similar microstructure (Figure 7). The same bismuth beads, apatite, and zincochromite crystals are included in a vitreous matrix.

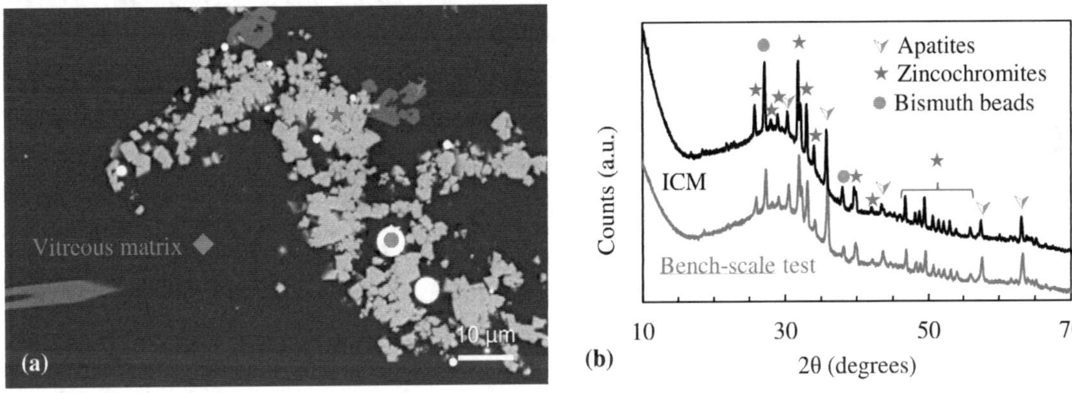

Figure 7. (a) SEM image and (b) XRD analysis of the wasteform produced using the ICM. The XRD results are compared to those of the bench-scale test.

7. Conclusions

The In-Can Melter trial conducted in the framework of the THERAMIN project demonstrated a successful process for the vitrification of ash generated from the incineration of organic waste from glove boxes in the nuclear industry. This study presents — in a simplified version — the scale up methodology used to demonstrate the feasibility of the thermal treatment of a given waste by a given process. Thus, the laboratory-scale tests assess the waste loading, operating temperature, mixture strategy and melted material quality. Next comes the bench-scale test focusing on crucible corrosion, volatility of some species and wasteform microstructure. Finally, the study concludes with a pilot-scale trial whose main outputs are overall corrosion and kinetic data, process performance, process and package descriptions.

A good waste loading of 50 wt.% was achieved in this first approach. The trial also made it possible to begin the technical investigation required for the processing of powdery solids into the can while avoiding dust emission: a temporary densification by pelletising was implemented.

The produced wasteform consists of a crystallised glass mainly composed of SiO_2, Na_2O, B_2O_3, Al_2O_3, and CaO. The term 'crystallised glass' refers to a vitreous matrix including crystals of apatite, zincochromite, and bismuth alloy. The crystals are distributed homogeneously in the wasteform.

The crystalline phases being durable, the durability of the wasteform is controlled by that of the vitreous matrix. The hydrolysis rate of this glass is relatively high because of its high content of B_2O_3 and Na_2O. Indeed, in the approach proposed in the framework of the THERAMIN project, no optimisation study of the glass additives was conducted. However, the 'classical' trends observed suggest that, in the long term, an alteration layer will form, leading to a decrease in the alteration rate.

IOP Conf. Series: Materials Science and Engineering **818** (2020) 012005 doi:10.1088/1757-899X/818/1/012005

Acknowledgements
This project has received funding from the Euratom research and training programme 2014–2018 under grant agreement No 755480. This paper reflects only the authors' views, and the European Commission is not responsible for any use that may be made of it.

The authors are grateful to Alain Artico, Carine Castaño, and Virginie Lemaitre for carrying out the tests, Thierry Blisson, Florian Emanuel, and Valérie Debono for the wasteform characterisation, and PRIME Verre Company for its technical assistance.

References

[1] ASTM International 2014 *Standard C1285-14. Standard Test Method for Setermining Chemical Durability of Nuclear, Hazardeous, and Mixed Waste Glasses and Multiphase Glass Ceramics: The Product Consistency Test (PCT)*

[2] Brunauer S, Emmett P H and Teller E 1938 Adsorption of gases in multimolecular layers *J. Am. Chem. Soc.* **60** 309–19

[3] Nicoleau E, Angeli F, Schuller S, Charpentier T, Jollivet P and Moskura M 2016 Rare-earth silicate crystallization in borosilicate glasses: Effect on structural and chemical durability properties *J. Non-Cryst. Solids* **438** 37–48

[4] Gin S, Abdelouas A, Criscenti L J, Ebert W L, Ferrand K, Geisler T, Harrison M T, Inagaki Y, Mitsui S, Mueller K T, Marra J C, Pantano C G, Pierce E M, Ryan J V, Schofield J M, Steefel C I and Vienna J D 2013 An international initiative on long-term behavior of high-level nuclear waste glass *Mater. Today* **16** 243–8

[5] Kaspar T C, Ryan J V, Pantano C G, Rice J, Trivelpiece C, Hyatt N C, Corkhill C L, Mann C, Hand R J, Kirkham M A, Crawford C L, Jantzen C M, Du J, Lu X, Harrison M T, Cushman C, Linford M R and Smith N J 2019 Physical and optical properties of the International Simple Glass *npj Mater. Degrad.* **3** 15

[6] Gin S, Beaudoux X, Angéli F, Jégou C and Godon N 2012 Effect of composition on the short-term and long-term dissolution rates of ten borosilicate glasses of increasing complexity from 3 to 30 oxides *J. Non-Cryst. Solids* **358** 2559–70

IOP Conf. Series: Materials Science and Engineering **818** (2020) 012006 doi:10.1088/1757-899X/818/1/012006

Plasma Technology to Recondition Radioactive Waste: Tests with Simulated Bitumen and Concrete in a Plasma Test Facility

J Deckers

Belgoprocess N.V. Gravenstraat 73, 2480 Dessel, Belgium

jan.deckers@belgoprocess.be

Abstract. The operation and maintenance of nuclear power plants, the non-nuclear fuel cycle, etc generate low-level radioactive waste which, along with the historical radioactive waste from past nuclear activities, needs to be treated and stored, awaiting final disposal. Plasma technology offers a very effective way of treating this waste with a high-volume reduction factor (VRF), free from organics, liquids and moisture, and meets without a doubt the acceptance criteria for safe storage and disposal. By means of a plasma beam of approximately 5000°C, the inorganic materials are melted into a glassy slag, containing most of the radioactive isotopes while the organic material is gasified, oxidized and purified in an off-gas cleaning system. First the paper describes the new full-scale Plasma Melting Facility (PMF) at the **Kozloduy Nuclear Power Plant in Bulgaria** which was taken into nuclear operation in May 2018. The plant has a capacity of 250 tons per year and the maximum contact dose rates of the incoming waste is 2mSv/h. Different mixtures of radioactive waste packed in 200l drums were successfully treated resulting in a glassy slag free from liquids and organic material with an important volume reduction factor (VRF). The Project was co-financed through a grant by Kozloduy International Decommissiong fund (KIDSF) administated by the EBRD through Bulgarian national funding. Plasma is a suitable technology for treatment of problematic waste or even reconditioning waste so Belgoprocess was contracted to do plasma tests with simulated conditioned waste types. One can do tests on a laboratory scale on smaller samples and torch capacities of e.g. 50kW but Belgoprocess wanted to do more realistic and reliable tests. So Belgoprocess contracted Phoenix Solutions Co who has a full-scope test facility equipped with a 1200kW plasma torch for full-scope treatment of simulated conditioned waste. For a first confidential contract simulated 200l (55 gallon) bitumen drums were treated. The drums contained different pucks of compacted waste such as rags, used filters, granulates, etc. The pucks were stacked in the 200l drums and subsequently embedded with bitumen. A total of 6 drums were treated in the plasma facility. For a second contract simulated homogeneous 200l (55 gallon) concrete drums with on the one hand concentrates and on the other hand spent resins were selected. A total of 6 drums with concrete and spent resins were treated and melted in the plasma testing facility. The paper describes the test facility, volume reduction factor (VRF) of different waste streams and most important parameters.

1. A proven technology: the new full-scale plasma facility at Kozloduy NPP

More and more there is a growing interest for the use of plasma technology. Common technologies to treat different waste types are sorting, supercompaction, drying and immobilisation grouting. But due to the increasingly higher requirements for acceptance criteria for final disposal of radioactive waste these technologies are often not sufficient. Even already conditioned drums with e.g. bitumen matrix don't comply with current acceptance criteria. The innovative plasma technology can treat these complex and problematic waste mixtures in one single process with a high-volume reduction factor (VRF), free from organics, liquids and moisture, and meeting the acceptance criteria for safe storage and disposal. Indeed, by means of a plasma beam of approximately 5000°C, the inorganic materials such as concrete debris, sand, inorganic granulates, insulation material and even asbestos are melted into an amorphous glassy slag, containing the concentrated radioactive isotopes, while the organic material is gasified and afterwards oxidized in an afterburner and purified in an off-gas cleaning system.

A new full-scale plasma facility was taken into nuclear operation at the Kozloduy Nuclear Power Plant site [1]. The plant was built by the Joint Venture Iberdrola Ingeniería y Construcción - Belgoprocess as an engineering, procurement and construction (EPC) contractor. The Project was co-financed through a grant by the Kozloduy International Decommissiong fund (KIDSF) administated by the EBRD and through Bulgarian national funding. The KIDSF is funded by the European Union as well by the contributors to the KIDSF – Austria, Belgium, France, Grece, Ireland, The Netherlands, Spain, United Kingdom and Switserland. The plasma facility is operated by the State Enterprise of Radioactive Waste in Bulgaria (SERAW).

The facility consists of a tilting plasma furnace equipped with a non-transferable torch of 500 kW as heat source and will treat 250 tons per year, spread over 40 operational weeks. The tilting furnace developed by the JV has been designed to pour the slag in a controlled way into a slag mould. This concept of the furnace with the fixed waste feeder, the off-gas equipment and the closed confinements around the slag pouring, prevents the escape of radioactive or hazardous gasses and particles into the work area and into the atmosphere, thereby improving the safety features of the plasma facility. Solid waste to be processed will be delivered in:

- Bags containing mainly organic waste
- Metallic 200 l drums containing pre-compacted organic waste and metal particles.
- Pucks resulting from the super compaction of metallic 200 l drums containing mixtures of concrete, wood and other organic material. Pucks have heights up to 40 cm

Typical specific radioactivity of the waste is about 5E5 Bq/kg and contact dose rate is lower than 2mSv/h. It is also foreseen to treat liquids and drummed spent ion exchange resins.

The incoming waste is transferred to a shredder and from there to the feeder tube. Close to the furnace, the feeder tube has a rotating connection so that the feeder tube of the shredder is fixed-mounted in relation to the tilting furnace. On the opposite side of the furnace the contaminated hot gasses with a temperature of 1300°C are diverted to the afterburner chamber.

The system processes mixtures of organic waste such as plastic and celluloses, with inorganic waste such as concrete, mineral insulation, glass and metals. Depending on the incoming waste composition, a glassy-like slag or a metal-like slag is obtained. When about 200 litres of slag are produced, the slag is poured into the slag mould. After pouring, about 50 litres of slag remains in the furnace and it is used as a thermal flywheel for the next waste batch. The remaining slag also forms a protection for the refractory against the high temperature of the plasma flame.

The tapping of slag into the slag mould is carried out in a confinement in order to prevent spreading of contamination in the normal work area and into the environment.

An overview of the plasma facility is given in Figure 1 and consists of the following main components:

- A robust dual shredder with extruder feeder with a nitrogen blanketing system
- A primary treatment chamber equipped with a non-transferable torch and a sealed slag collection chamber
- A secondary combustion chamber in which the syngas is mixed with excess air to complete oxidation to primary combustion components
- A boiler to cool the off-gasses
- Off-gas filtration and radiological purification, consisting of a bag-house with bag filters and high efficiency particulate air (HEPA) redundant filters

- A wet off-gas scrubbing system, consisting of a quench tower and a counter current scrubbing tower to remove HCl and SO_2
- Redundant flue gas extraction fans for keeping continuous negative pressure into the entire system and evacuating the flue gasses
- A DENOX system provided with a catalyst in order to reduce the NO_x
- At the end a continuous emission monitoring (CEM) and radiomonitoring system is also foreseen

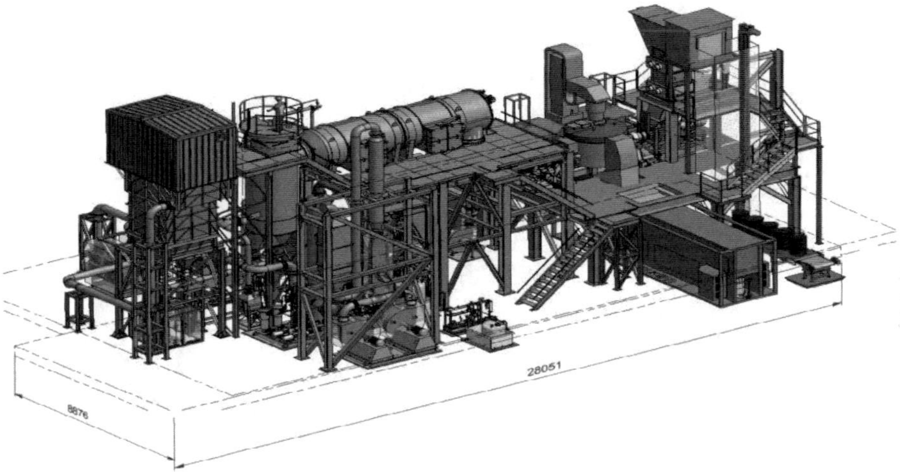

Figure 1. Overview of the Plasma Facility.

2. The PSC Plasma Test Facility

Nowadays it happens that future potential users of the plasma technology want to do tests on simulated conditioned waste which has to be reconditioned. One can do tests on a laboratory scale on smaller samples and torch capacities of e.g. 50kW but Belgoprocess wanted to do more realistic and reliable tests. So Belgoprocess contracted and worked together with Phoenix Solutions Co (PSC) who has a full-scope test facility equipped with a 1200kW plasma torch for treatment of simulated conditioned waste. By means of the full-scope plasma tests the beneficiary can observe that plasma treatment of these waste types is realistic, evaluate the obtained VRF and quality check the obtained glassy slag. If needed the beneficiary can organise by himself additional composition analyses, leachability and much more tests on the obtained slag.

The different systems that constitute the PSC arc plasma melting/gasification facility used for the above tests are the following:

- Drum feeder
- Plasma chamber
- Off-gas system
- Slag collection system

Through the feeding system, whole single 200l drums can be fed directly into the plasma treatment chamber (see Figure 2 below). The drum feeder is separated by a sealed thermally-protected slide gate. The plasma chamber is a high temperature (refractory-lined with 1500°C max temperature) tilting furnace. The fully-articulating plasma torch of up to 1200 kW is a non-transferred arc plasma torch.

After plasma treatment, the flue gases entered into the off-gas system to remove the chemical species to an acceptable level in the process obtaining flue gases practically dust free. The off-gas system configuration of the plasma facility at HTF (Hutchinson Test Facility) consisted of cyclone for large particulate removal and capture, secondary combustion chamber, eductor system for plasma chamber suction control and a bag house filtration for fine particulate capture.

The hot slag was poured into a containment vessel that allowed the cooling to form an inert glassy slag of minimum volume. After the cooling process the slag was transferred into different 200l drums by keeping traceability of the original test drum. At the end all the collected slag was available for the beneficiary.

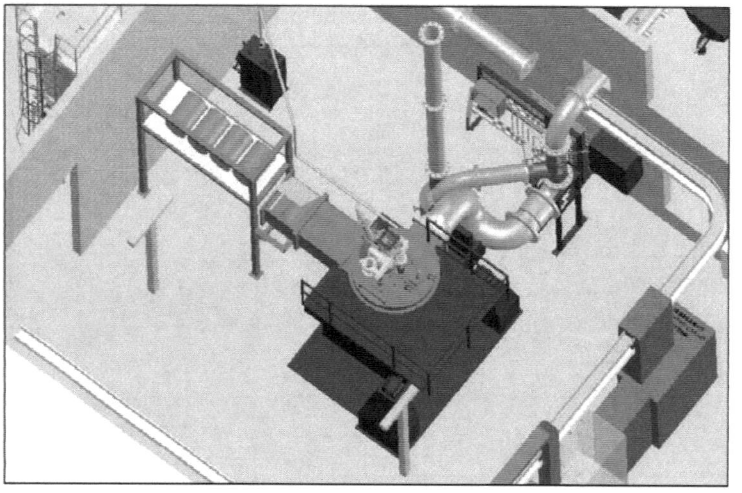

Figure 2. Feeding System for Plasma Processor at PSC's Hutchinson Test Facility.

For a first contract simulated homogeneous 200l (55 gallon) concrete drums with on the one hand concentrates and on the other hand spent resins were selected. A total of 6 concrete drums as conditioning matrix with concentrates and spent resins were treated and melted in the plasma testing facility. For a second confidential contract simulated 200l (55 gallon) bitumen drums were treated. The drums contained different pucks of compacted waste such as rags, used filters, granulates, etc. The pucks were stacked in the 200l drums and subsequently embedded with bitumen. A total of 6 drums were treated in the plasma facility.

3. Test with Concrete Waste Types

First a general control of the test facility was carried out for controlling proper functioning of all the equipment. After heating up, 6 drums were processed during a 3 working days test period. The 4 first drums (simuli 1-4) contained the homogeneous conditioned concentrates in a concrete matrix. Typical weights for filled up 200l drums was 420kg and contained 68l of concentrate solution with a density of 1,35kg/l. The composition of the concentrates was mainly different types of sodium salts (borates, sulphates, chlorides, nitrates). Drum identification simuli 2 was only filled up for half of the volume so the weight is only 180kg. The last 2 drums (R1-R2) contained the homogeneous conditioned cationic and anionic resin in a concrete matrix. Weights were almost 400kg per 200l drum and contained 68l of resins. Figures 3 to 6 illustrate the different stages of treatment of a drum with concentrates.

Figure 3 illustrates the drum with removed lid with the homogeneous conditioned concentrates in a concrete matrix. For the plasma melting the whole drum inclusive lid and bolts for closure are fed to the furnace. Figure 4 shows the pouring of the hot slag at around 1450°C at the end of the melting process.

IOP Conf. Series: Materials Science and Engineering **818** (2020) 012006 doi:10.1088/1757-899X/818/1/012006

Figure 3. Concentrates in concrete.

Figure 4. Pouring of hot slag.

Figure 5 illustrates the amount of collected slag of 1 drum. After cooling down, the volume of the slag poured into the calibrated receptacle was determined and subsequently the VRF was calculated. Indeed the volume is less than the original drums as the bounded water with cement is evaporated. At the end a more dense slag of around 3kg/l is obtained in comparison with the original density of about 2.2kg/l. Figure 6 gives a representative picture of the obtained slag. Visually it can be seen that the final product is similar to a vitrification process.

Figure 5. Collected slag just after pouring.

Figure 6. Representative chunk of slag

Table 1, below, gives an overview of the concrete drums with original weights, applied power and VRF.

Table 1. Result with VRF

Drum Identification	Drum Weight (kg)	Torch Power, (kW)	VRF
Simuli 1	445	1200	1.3
Simuli 2	180	1200	1.3
Simuli 3	414	1200	1.4
Simuli 4	437	1200	1.3
R1 IRN 77	392	1200	1.6
R2 IRN 78	368	1200	2.0

4. Test with Bitumen Waste Type

For a second contract simulated 200l (55 gallon) bitumen drums were treated. The drums (see Figure 7) contained different pucks of compacted waste such as rags, used filters, granulates, etc. The pucks were stacked in the 200l drums and subsequently embedded with bitumen. A total of 6 drums were treated in the plasma facility.

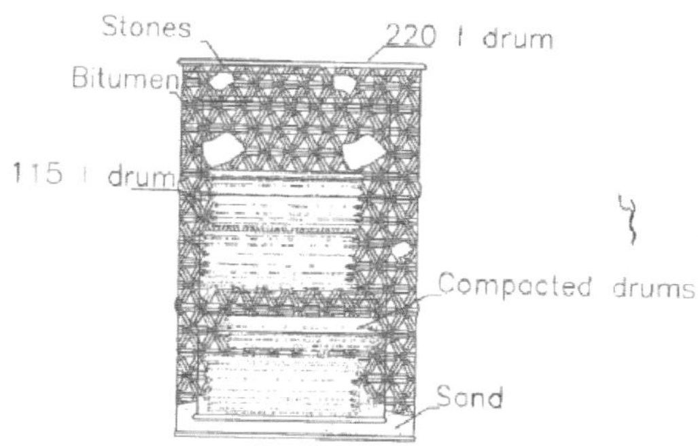

Figure 7. Simulated 200l bitumen drum (SBD) with compacted 115l drums

Each drum contains 140kg of bitumen which corresponds with a power production of 1600kW. When these drum are loaded into the hot furnace as is, a high instantaneous gas production will be developed and subsequently the oxidiser and off gas cannot follow in order to completely oxidise the flue gasses to primary products such as CO_2, H_2O, HCl and SO_2.

Therefore the drums are fragmented one by one in another facility. This was done in a KOMAR shredder of 60" by 40" and 150KW power. The shredded material was then collected into different drums taking into account full traceability. The simulated bitumen drums (SBD) had a weight of about 250kg containing 140kg bitumen as conditioned matrix. Six of these drums were shredded or fragmented and collected in 30 or 55 gallon steel drums. After shredding each original SBD resulted in 4 drums. All the drums (24 drums) were then packed and transported to the Hutchinson Test Facility.

During a 3-day test period all the 24 drums resulting from 6 original drums were treated into the plasma test facility during day time. The drums are introduced into the hot furnace one by one with the torch off. During about 30 minutes the drums containing the high amounts of bitumen were gasified. The unburned gasses and soot were then sent to the oxidizer with addition of maximum amount of air in order to obtain complete combustion and low CO content of the flue gasses (less than 50mg/Nm³). After that the plasma torch was taken into operation at a power of a 700kW for about 20 to 30 minutes for melting the inorganic materials such as granulates and steel.

At the end all the slag was collected in one receptacle (see Figure 8) in 2 pours at temperatures of about 1400°C resulting in 235l slag. One could observe that slag contains a lot of Fe_2O_3 (magnetite) coming mainly from melting and oxidizing the different steel drums containing the fragmented bitumen. Also the slag was free of any organic material.

Figure 8. Collected total amount 235l slag in a 272l receptacle.

In fact, during fragmentation the material was repacked into 24 additional steel drums. The empty 24 steel drums have a total weight of 316kg which result in 436 kg Fe_2O_3. Taking into account the density of 5.18kg/l for magnetite additional 84l litres were added to the slag. At the end the conservative and realistic VRF can be calculated. The conservative VRF can be defined as the ratio between the six incoming drums (6*217l) and the total volume of the receptacle (272l) which gives a VRF of 4,8.

The net VRF can be defined also as the ratio between the six incoming drums (6*217l) and the net volume of slag minus the slag coming from the additional steel drums (235l-84l) which resulted in a net VRF of 8,6. Indeed in reality shredded drums will be sent directly to the plasma furnace without necessary repacking.

5. Conclusions

- Plasma is a proven technology which can treat problematic waste types resulting in a slag free of organics and similar to the product of vitrification.
- Purpose of both tests namely to recondition simulated conditioned waste was successfully carried out: melting of simulated drums, pouring of the molten slag and collection of the solidified slag material.
- Volume reduction factor (VRF) for simulated concrete drums containing concentrates can be set at 1,3 and for concrete with the spent resins at 1,5.
- Volume reduction factor (VRF) for simulated bitumen drums (SBD) containing different compacted pucks has a net VRF of at least 8.

Acknowledgements
Thanks to Mr Garry Hanus from Phoenix Solutions Co (USA) for the contribution and review of this paper

References
[1] Deckers J 2019 Taking in nuclear operation of the Plasma Facility for industrial treatment of radioactive waste *Proc. of the WM Symp.* 2019 Phoenix USA.

IOP Conf. Series: Materials Science and Engineering **818** (2020) 012007 doi:10.1088/1757-899X/818/1/012007

Gasification-based thermal treatment of Low and Intermediate Level Waste containing organic matter

M Nieminen*, J Laatikainen-Luntama and M Olin

VTT Technical Research Centre of Finland Ltd., P.O.Box 1000, FI-02044 VTT, Finland

*matti.nieminen@vtt.fi

Abstract. VTT has developed a gasification-based technology for processing low- and intermediate-level waste (LILW) rich in organic matter. VTT's process is based on well-controlled thermal fluidised-bed gasification followed by efficient gas cleaning, gas conditioning/oxidation, wet scrubbing of the oxidised gas (flue gas), and finally, filtering through a high efficiency particulate air (HEPA) filter, which acts as a backup cleaning system. In the case of an organic ion exchange resin (IXR) the primary product from thermal gasification is fine filter dust. In addition, the process produces some bottom ash, which consists mainly of the bed material. The wet scrubber liquid may also contain some activity. Demonstration trials were carried out using unspent organic IXR containing a small amount of added stable Cs in order to simulate Cs content in spent IXR. Gasification tests confirmed the capability of the process to remove organic matter from the IXR and clean the resulting off-gas as required. Simulated waste IXR was reduced to approximately 1 wt-% of its original mass before immobilisation. Filter dust and bottom ash have to be immobilised before final disposal. VTT has previously developed an advanced immobilisation technology based on geopolymerisation and this has been applied to samples from the THERAMIN test trials. VTT's technology has been designed as a compact process, which can be operated at the nuclear power plant site. Until now, all development and verification test trials have used simulated waste materials. The next step is the demonstration of the process using real radioactive waste.

1. Introduction

VTT has, over several decades, developed gasification and gas cleaning technologies for different applications and different feeds [1, 2, 3]. The primary focus has been on production of fuel gas or ultra clean synthesis gas from high quality fuels, but later in the 1990s development also focused on using different wastes and industrial by-products containing harmful contaminants that need to be removed from the resulting gas; this experience has enabled the development of thermal gasification technology for treatment of low- and intermediate-level radioactive waste (LILW).

The gasification-based process developed by VTT is based on well-controlled thermal fluidised bed gasification followed by efficient gas cleaning, gas oxidation and wet scrubbing of the oxidised gas. The process has been verified to be a suitable thermal treatment method for organic ion exchange resin (IXR). However, this technology can also be applied to any organic waste streams, including liquid organics. Removal of organic matter by thermal gasification results in significant volume reduction, which was demonstrated successfully in the THERAMIN project. The main final residue is filter dust, which requires further processing in order to immobilise it and enable safe disposal. Immobilisation can be achieved by cementation. However, VTT has developed an advanced immobilisation technology based on geopolymers, which enables high waste loading and has shown lower leachability characteristics than Ordinary Portland Cement immobilisation. Geopolymerisation was used to immobilise the samples from demonstration test trials carried out in the THERAMIN project [4].

2. Gasification based LILW treatment technology

Thermal gasification is a mature technology that has been used for almost 200 years to produce energy from carbon-containing fuels. Gasification is a partial oxidation process. Conditions in the gasification reactor are reducing, and the gas produced still contains combustible compounds. Some oxygen is needed to produce the heat required for thermal gasification reactions, but the gas atmosphere is reducing which has an impact on inorganic chemistry in the reactor.

The major radionuclides in organic LILW are typically metals such as caesium and nickel, or carbon (C-14). Through gas cleaning, all solid material can be removed easily prior to gas combustion. Gasification temperature is significantly lower than high temperature vitrification and thus only a small proportion of some metals will be vaporised. These vaporised metals can be solidified by cooling the gas prior to particulate removal based on efficient high-temperature filters, and thus metals can be removed together with other solids. In addition, in radioactive waste treatment, the flue gases are cleaned after gas oxidation (combustion) by filtering through a High Efficiency Particulate Air (HEPA) filter.

VTT has Bubbling Fluidised Bed (BFB) and Circulating Fluidised Bed (CFB) gasification test rigs suitable for thermal gasification of organic IXR or other organic LILW. The demonstration presented in this paper was carried out using an atmospheric pressure pilot-scale CFB gasification test rig (Figure 1) at VTT's Piloting Centre. The bed diameter of the reactor was 102 mm and the freeboard diameter was 150 mm. Electrical heaters were situated around the reactor and gas pipes to compensate for heat loss. The test rig was equipped with versatile fuel feeders. In addition, the test rig was equipped with a separate feeder for the bed material/additive. The fuel feeding port was located above the dense bed. The particles from the recycling cyclone were introduced close to the air distributor. Electrically preheated air was used as a gasification agent.

The gasification product gas was cleaned in the recycling cyclone followed by a gas cooler and a high-temperature filter unit. The filter unit consisted of 12 metal candle elements, which were approx. 60 mm in diameter and 1000 mm in length. Nitrogen was used as a filter pulsing gas (back pulse cleaning). The role of the filter unit was to remove particles from the product gas. Easily volatile compounds condense into solid particles when the gas cools and when the filtering temperature is low enough, and harmful compounds (e.g. Cs in these test trials) are thus removed from the process along with the filter dust.

The test rig was equipped with an automatic data acquisition system to measure temperatures and pressures at different locations. The flow rates of gasification agents and purging nitrogen were continuously measured and recorded. The fuel feed rate was controlled adjusting the feed rate of IXR on the basis of the screw calibration curve, and the rotation speed of the metering screw was continuously measured. In addition, feedstock (IXR) batches were always weighed when loading into the feeding hopper in order to check the actual feedstock consumption after the test trial. The main gas components CO, CO_2, H_2, CH_4 and O_2 are monitored by on-line analysers. All mass streams were measured except product gas, which was calculated based on the other mass streams and the product gas composition. All solid residues (bottom ash and filter dust) were collected and analysed.

IOP Conf. Series: Materials Science and Engineering **818** (2020) 012007 doi:10.1088/1757-899X/818/1/012007

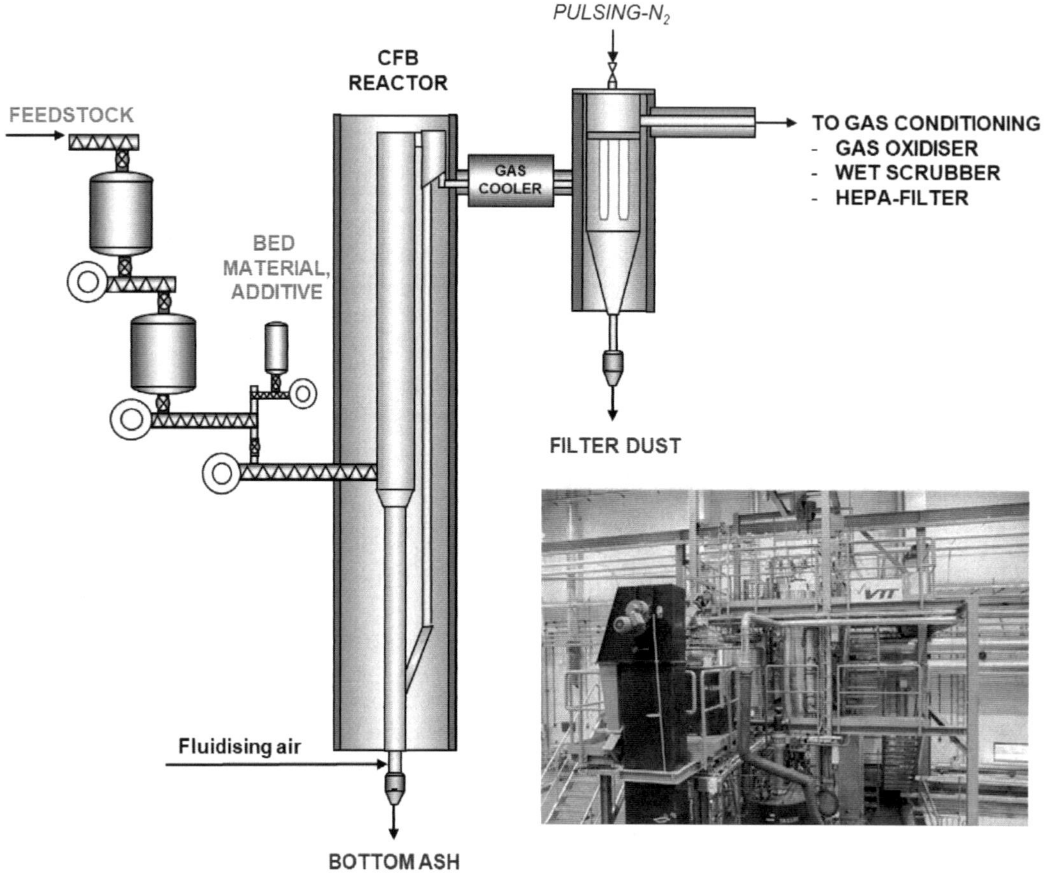

Figure 1. Pilot-scale Circulating Fluidised Bed (CFB) gasification test rig.

3. Demonstration trials

Demonstration trials were carried out in the pilot-scale CFB gasifier using IXR as a feed. The feed was impregnated with CsCl in order to simulate Cs content in a real spent IXR. A total of 325 kg of organic IXR was treated over the course of three trial days. The total duration of the trial was 26.5 hours. The average gasification temperature was 885-915°C, and the filtering temperature varied between 415 and 450°C. A total of 18 kg of inert Al_2O_3 (of particle size 0.18-0.25 mm) was used as a bed material. The dry gas composition prior to oxidation during the second test trial day is presented in Figure 2.

The success of the trial is primarily assessed by determining mass balances for the main elements involved in the reactions. The mass balances for the trials were calculated based on the average values of the measured data (including ash and ultimate analysis of the IXR and filter dust, dry gas composition as well as tar content in the dry product gas). The carbon conversion calculation is based on the carbon mass balance and it describes how efficiently input carbon is converted to gas. The carbon conversion to gas and tars was over 98.3 %, which means that the removal of the organic material from the IXR was successful. The product gas was dilute because the equivalence air ratio was kept deliberately rather high (0.49-0.54) since the main target was to remove the organic matter from the feedstock (IXR), not energy production. The mass and volume of waste to be disposed was significantly reduced: a total of 325 kg of organic IXR was gasified during three test trials, and 3.3 kg of filter dust was produced during the whole feedstock feeding time, which represents about 1 wt-% (1.4 wt-% dry) of the fed IXR.

The demonstration trials aimed to verify the effectiveness of gasification-based technology to reduce the volume of organic IXR and to produce treated waste (solid residue) for further characterisation tests. The product of thermal gasification was powder residues and therefore immobilisation was required to

IOP Conf. Series: Materials Science and Engineering **818** (2020) 012007 doi:10.1088/1757-899X/818/1/012007

enable final disposal. The solid residue can be immobilised by applying different technologies; in the THERAMIN project, geopolymerisation was used. The methodology for this process was developed during previous VTT projects and so far the results have been significantly better than immobilisation using Ordinary Portland Cement (OPC) [4]. Geopolymerisation enables high waste loading and simultaneously leachability of radionuclides is significantly lower compared to immobilisation by OPC.

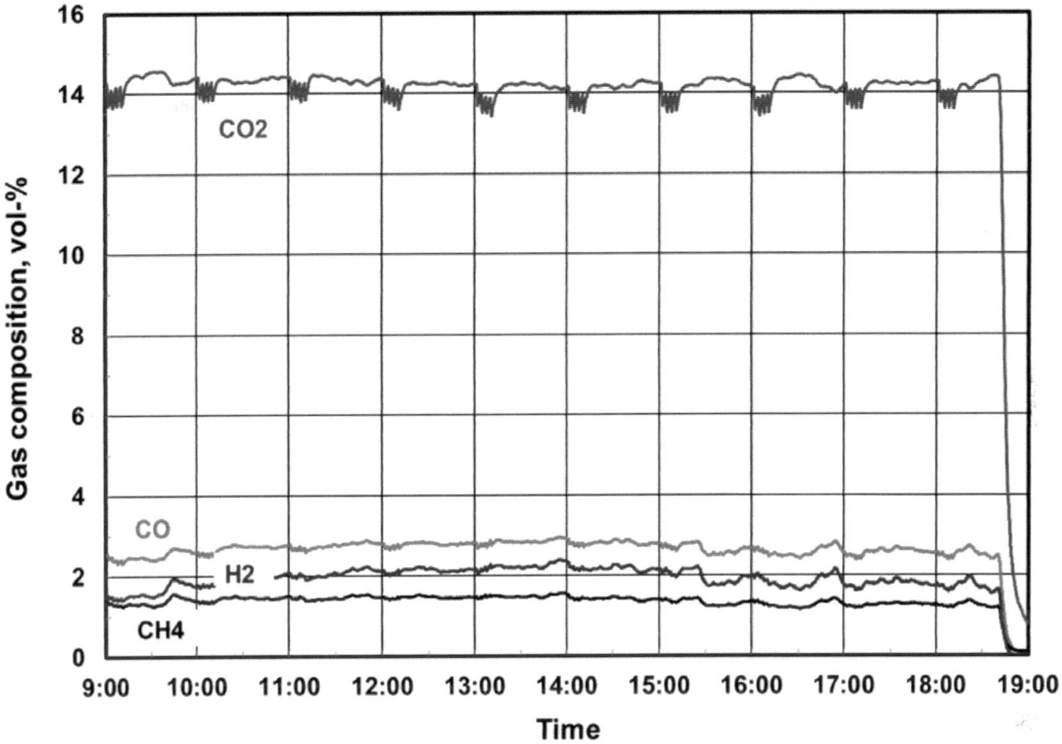

Figure 2. Main dry gas composition during the second trial day (the main gas component is nitrogen, which is not shown in the figure).

4. Conclusions

The performance of the gasification-based thermal treatment of organic ion exchange resins developed by VTT was demonstrated in the THERAMIN project. The treated waste (solid residue from the gasification treatment) was further immobilised by geopolymerisation. Demonstration trials including geopolymerisation verified that the developed thermal gasification, gas conditioning and geopolymerisation of treated waste effectively consigns all metallic radionuclides within the ash products, reduces waste volume significantly, does not cause any off-gas emissions, and demonstrates that the ashes can be immobilised effectively through geopolymerisation. The development will be continued further, in particular targeting demonstration with active wastes from nuclear power plants.

Acknowledgments

The project has received funding from the Euratom research and training programme 2014-2018 under grant agreement No 755480. This paper reflects only the authors' views, and the European Commission is not responsible for any use that may be made of it.

References

[1] Kurkela E, Moilanen A, Nieminen M 1999 CFB gasification of biomass residues for co-combustion in large utility boilers: Studies on ash control and gas cleaning *Power Production from Biomass III: Gasification and Pyrolysis R&D&D for Industry. Espoo: VTT Technical Research Centre of Finland* (VTT Symposium; No. 192) pp 213-228

[2] Kurkela E 2002 Gasification of waste-derived fuels: R&D activities at VTT *Power production from waste and biomass IV: Advanced concepts and technologies. Espoo: VTT Technical Research Centre of Finland* (VTT Symposium; No. 222) pp 267-276

[3] Nieminen M, Suomalainen M, Mäkinen T 2006 Gasification of shredder residue. Espoo: VTT Technical Research Centre of Finland (VTT Tiedotteita - Meddelanden - Research Notes; No. 2344) p. 52

[4] Vehmas T, Myllykylä E, Nieminen M, Laatikainen-Luntama J, Leivo M and Olin M 2020 Geopolymerisation of gasified ion-exchange resins, mechanical properties and short-term leachning studies, *IOP Conf. Ser.: Mater. Sci. Eng* same issue

IOP Conf. Series: Materials Science and Engineering **818** (2020) 012008 doi:10.1088/1757-899X/818/1/012008

Pyrolysis of Radioactive Spent Resins in the PRIME Installation

J Hansen[1] and J Deckers[2]
[1] Montair Process Technology B.V.
[2] Belgoprocess N.V.

Email: jurgen.hansen@belgoprocess.be

Abstract. Montair and Belgoprocess developed the pilot PRIME (Pyrolysis Resins In Mobile Electric) Installation for treatment of spent resins in one single reactor without pre-treatment and without intermixing of other waste, allowing for easy traceability and isotopic characterization of the end product. Both low level waste (LLW, < 2 mSv/h) and medium level waste (MLW, > 2 mSv/h) can be treated in the installation. The PRIME pilot installation is designed to treat 2 batches of 100 litre of saturated ion exchange resins over a period of 24 hours. The pyrolysis reactor contains a special design mechanical stirrer and can operate at a temperature of up to 600°C under inert conditions. As a result, the organic molecules will decompose at an accelerated rate until only carbon-rich and mineralised product free of hydrocarbons remains. Gases that are released during drying and pyrolysis are treated in a thermal oxidizer, which uses an innovative, electrically heated design instead of a traditional natural gas- or fuel oil secondary combustion chamber design. After passing through the thermal oxidizer, the gases are cooled down to saturation temperature and treated in the wet scrubbing system. An induced draft fan keeps the system under negative pressure to prevent the possibility of any off-gas leakage out of the system. The pyrolysing reactor and oxidiser are fully electrically powered which creates also a higher safety level. A series of tests were performed successfully with wet cationic resins, anionic resins and a mixture of both. The results of these tests are described in the presentation. For intellectual property protection, the system and process of pyrolysis of organic waste and the system of thermal oxidation of pyrolysis gases described in this paper are patent pending.

1. Introduction

Internationally, handling of the continuously generated waste stream of radioactively contaminated ion exchange resins remains a critical issue [1], [2]. While various treatment installations are operational, a true mineralization (conversion to unreactive inorganic products) of spent ion exchange resins is not obvious due to the resins' activity and isotopic composition. Mineralization is desirable however, since it reduces waste volume and eliminates the possibility for further decomposition of organic material during storage. Stored and untreated spent resins may provoke chemical reactions from or between the organic resin, the container/drum, filler material and concrete. A chemical reaction can affect the storage quality of the products and can compromise the container, resulting in leaking storage units. Treatment of such spent resins must therefore meet severe criteria for the organic waste content.

The problem with incineration is that it is an uncontrolled thermal process where the temperature in the waste can easily rise above 1.000°C. The high temperatures obtained during incineration will lead to carry-over of isotopes such as caesium, while the quick evolution of gasses will lead to carry-over of contaminated dust particles. An additional problem that occurs during incineration is that the resin beads will clog together when they are fed in packages, merging together into large chunks. Clearly, this clogging together of the resin beads negatively impacts the effectiveness of treatment, as it inhibits heat- and mass transfer.

Pyrolysis of organic waste under nitrogen atmosphere on the other hand is a controlled process. [3] The temperature in the waste can be limited, preventing carry-over of radioactive isotopes. The controlled conditions also prevent the undesired clogging together of individual resin beads. Currently, pyrolysis is already a "proven technology" for specific applications in the nuclear industry, such as for pyrolysis of tributyl phosphate (TBP)-containing liquids and/or plutonium-containing organic products. [4]

2. The Prime – Installation Working Principles

The PRIME pilot installation is designed to treat 2 batches of 100 L of saturated ion exchange resins over a period of 24 hours. A dosing unit with screw feeder is used for the dosing of the saturated ion exchange resins to the pyrolysis reactor. The pyrolysis reactor contains a special design mechanical stirrer and can operate at a temperature of up to 600°C under inert conditions. As a result, the organic molecules will decompose at an accelerated rate until only the mineralized material remains. The lower temperature and the absence of oxygen also prevent the breakdown of ion exchange resins into (powdery) ash, which is a considerable advantage during further handling.

Gases that are released during drying and pyrolysis are treated in a thermal oxidizer, which uses an innovative, electrically heated design instead of a traditional natural gas- or fuel oil secondary combustion chamber design. After passing through the thermal oxidizer, the gases are cooled down to saturation temperature and treated in the wet scrubbing system. An induced draft fan keeps the system under negative pressure to prevent the possibility of any off-gas leakage out of the system.

A series of tests were performed successfully with wet cationic resins, anionic resins and a mixture of both. The results of these tests are described in the presentation.

For intellectual property protection, the system and process of pyrolysis of organic waste and the system of thermal oxidation of pyrolysis gases described in this paper are patent pending.

3. Research and Development

Montair and Belgoprocess conducted an R&D project to investigate the feasibility of mineralizing ion exchange resins through a single process of drying and pyrolysis – a simple and straightforward process. To reduce the installation size and simplify possible dismantling afterwards, the installation size has been designed on the footprint of a 20-foot container. This small footprint also makes the installation mobile. To reach these goals, an innovative compact electrically heated oxidizer for the off-gas was designed, built and tested.

The pilot plant was designed, constructed and functionally tested, see Figure 1. For experimental testing, the pilot plant is designed to treat 100 L wet cationic resins or anionic resins or a mixture of both. The goal of the testing was to investigate the effectiveness of treatment of the ion exchange resins, to characterize the end product and to verify that no clogging occurred during treatment. In addition, the functionality of the dosing unit with screw feeder and the electrically heated oxidizer were tested.

The results of this pilot plant will be used to realize a mobile commercial installation for the treatment of radioactive spent ion exchange resins from nuclear power plants.

Figure 1. PRIME Pilot Installation, dimensions 2.3 x 5.7 x 3.8 m (w x l x h).

3.1. Ion exchange resins

The PRIME pilot installation is designed to process ion exchange resins. Ion exchange resins consist of a cross-linked matrix consisting of a styrene and divinylbenzene co-polymer. The resins can bind ions using specific functional groups. The cationic ion exchange resins contain sulfonate groups ($R-SO_3^-$), while the anionic ion exchange resins contain quaternary ammonium groups ($R-N(CH_3)_3^+$).

- Cationic ion exchange resins
  Molecular formula: $C_{25}H_{25}S_3O9$
- Anionic ion exchange resins
  Molecular formula: $C_{34}H_{52}N_3O_3$.

4. The Prime Pilot Installation Design

The PRIME pilot installation is easy to operate with the support of the fully automated control system. The PRIME pilot installation was operated by one operator during all tests.

The PRIME pilot installation is based on the treatment of 100 L of saturated ion exchange resins over a period of 24 hours. A dosing unit with screw feeder is used for the dosing of the saturated ion exchange resins to the pyrolysis reactor. The pyrolysis reactor contains a special design mechanical stirrer and can operate at a temperature of up to 600°C under inert conditions. Gases that are released during drying and pyrolysis are treated in a thermal oxidizer which uses an innovative, electrically heated design instead of a traditional natural gas- or fuel oil secondary combustion chamber design. After passing through the thermal oxidizer, the gases are cooled down to saturation temperature and treated in the wet scrubbing system. An induced draft fan keeps the system under negative pressure to prevent the possibility of any off-gas leakage out of the system. The flow diagram of the pilot plant is shown in Figure 2 below.

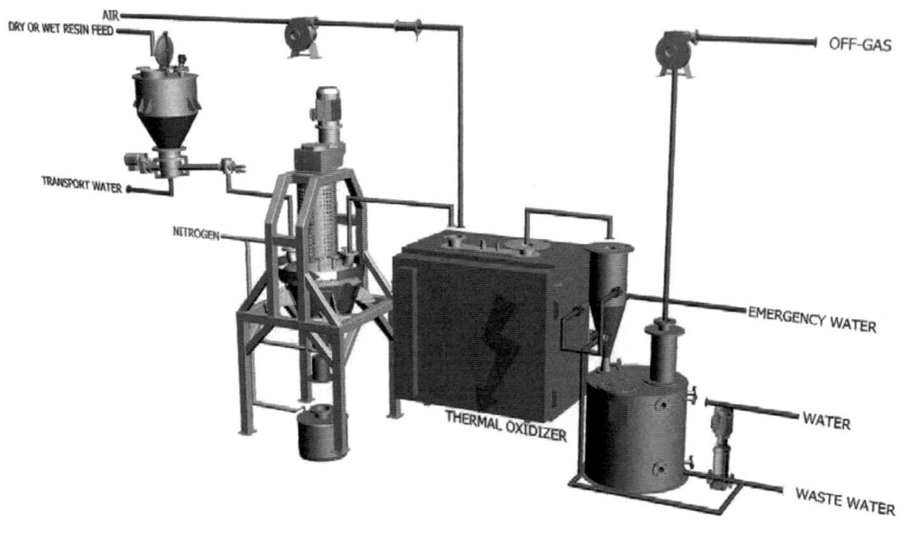

Figure 2. Flow diagram of the pilot plant based on a 20 ft container footprint.

4.1. Safe design principle

The PRIME pilot installation is designed with only electric heating as heating source, even thermal oils are excluded. As no flammable fuel sources such as oil or natural gas are used, there is no need for storage tanks and piping to the system. For a mobile system this is a huge advantage, as it allows the system to be placed at many more locations. No specific safety studies for storage tanks and transfer for fuel to the facility are needed.

The thermal oxidizer is designed to heat up using electrical power and control the internal temperature at the minimum 850°C, assuring the minimum 2s residence time of the off-gases in the presence of at least 6% oxygen by volume as in a conventional thermal oxidizer to ensure good oxidation of the pyrolysis gases.

Since no extra fuel is introduced for heating the off-gases by means of a burner, another big advantage is that no additional combustion gases are introduced by operating the burner. As a result, the new concept for the electrical oxidizer and the design of the other off-gas systems can be significantly smaller due to the reduced off-gas volume.

In the event of electrical power failure, the complete installation heating will be shut-down. The pyrolysis process in the reactor will stop automatically and the remaining pyrolysis gases will be oxidized in the oxidizer due to the accumulated residential heat of the bricks and excess of oxygen. Off-gas extraction fans (one e.g. as standby) as well as the pre- and HEPA filters can be operated in parallel as common practice for nuclear process installations.

4.2. Pyrolysis reactor

The PRIME pilot installation pyrolysis reactor has a useful volume of 100 L, with a design temperature of 600°C and is equipped with a special design mechanical stirrer on a central axis, see Figure 3. A resin inlet on the top connects the reactor to the dosing unit and a pyrolysis gases discharge connection at the top of the reactor connects the reactor to the thermal oxidizer. A discharge connection at the bottom of the reactor is connected to a receiver drum for the resins after treatment. Connections on the bottom and the top allow for nitrogen blanketing, ensuring an inert atmosphere. The inlet and discharge for the resins can be closed with pneumatically actuated slide valves. The pyrolysis reactor is heated by ceramic heating elements around the conical part of the reactor.

IOP Conf. Series: Materials Science and Engineering **818** (2020) 012008 doi:10.1088/1757-899X/818/1/012008

Although the installation does not make use of external flammable fuels, the pyrolysis gases themselves must also be considered as a flammable mixture when oxygen is introduced at high temperature. The pyrolysis gases cannot leak out of the installation, since the installation is kept at negative pressure by the extraction fans.

To prevent air leaking into the reactor section, before the thermal oxidizer, all possible leakages are sealed and/or flushed with nitrogen. On top of this, before start-up a leakage test is performed to ensure leak tightness.

Figure 3. Dosing unit (top left) and pyrolysis reactor (front). Dimensions approximately 1 x 0.1 x 1.3 m (w x l x h).

4.3. Dosing unit

Loading of the ion exchange resins into the reactor is controlled by a dosing unit with screw feeder, see Figure 3. The dosing unit contains a hopper to hold the ion exchange resins. This hopper can either be filled by using a pump or by manual loading. The saturated ion exchange resins are typically handled in 20- 40% free water, since the wet resins will not transfer well. The bottom of the hopper is fitted with a sieve to recover this free water, also known as transport water. A connection for nitrogen allows the oxygen to be purged from the ion exchange resins before loading into the reactor.

The loading of the ion exchange resins into the reactor is done with the screw feeder. The dosing can be set at a high rate, to quickly fill the reactor for batch operation, or at a low rate, for treatment of the ion exchange resins in a continuous process.

4.4. Thermal oxidizer

In the thermal oxidizer, the pyrolysis gases released during treatment of the ion exchange resins are heated electrically to 850°C at a minimum residence time of 2s, in the presence of at least 6% volume of oxygen. This ensures that over 99.9% of all volatile organic compounds (VOC's) are combusted. Traditionally, a natural gas or fuel oil burner is used for heating the thermal oxidizer (afterburner). However, the combustion products that are generated by the burner increase the size of the system considerably. Also, the burner flame prohibits a design for the thermal oxidizer that is not straight, making a compact design difficult to realize.

4.5. Wet scrubbing system

The hot off-gases from the thermal oxidizer are passed through an injector-venturi scrubber for further treatment, see Figure 4. In the injector-venturi, water is injected into the off-gases at high pressure. The off-gases are then forced through the narrow venturi throat section, greatly increasing flow velocity and atomizing the water to create a finely dispersed mist. At these conditions, particles and pollutants are quickly captured by the water droplets. In addition, the evaporating water cools down the hot off-gases till saturation temperature. After treatment, the cleaned off-gases are released into the atmosphere by use of an induced draft fan.

Figure 4. Thermal Oxidizer (left) and Wet Scrubbing System (front). Dimensions approximately 2 x 3.4 x 3 m (w x l x h).

Important note: for R&D purposes, a much-simplified design for the off-gas treatment system is used to save costs. Treatment of contaminated off-gases is a core business for Montair and will be designed when required.

5. Discussion of Experimental Results

The reactor was loaded with 100 L cationic saturated ion exchange resins with 6.8 % free water, containing 97.5 kg of wet resins. The resins were treated over 10 hours, at a maximum temperature of 550°C and under inert nitrogen atmosphere.

After drying and pyrolysis treatment, 13.2 kg of ashes were recovered from the 97.5 kg of wet cationic resins loaded into the reactor, a mass reduction of 7.4. A volume of 14.7 L of ashes was recovered from the 89.8 L of wet cationic resins loaded into the reactor. The volume reduction factor obtained was 6.1. A comparison by volume of the wet resins, dry resins and treated resins are shown in Figure 5. The wet resins are dark brown compared to the light brown colour of the dry resins. The wet resin beads are larger in size than the dry resin beads due to the absorbed water. The resins treated by pyrolysis are grey coloured and are smaller than the dry resins.

Figure 5. Wet ion exchange resins in water (left), dry ion exchange resins (middle) and residue after pyrolysis treatment (right).

The general mass balance for the pyrolysis process is shown in Table 1. The estimate of the total amount of gas released is 52.6 kg.

Table 1. General mass balance.

Wet resins with free water *(loaded)*	104.6	*kg*
- *Free water*	7.1	*kg*
Wet resins	97.5	*kg*
- *Absorbed water (calculated estimate)*	31.7	*kg*
Dry resins	65.8	*kg*
- *Residue recovered*	13.2	*kg*
Gas released over 4.5 hours	52.6	*kg*
Average mass flow	11.7	*kg/h*
Average gas flow after oxidation	18.0	*Nm^3/h**

*Nm^3: standard volume for gases at a pressure of 1 atmosphere and at a temperature of 0°C)

5.1. Pyrolysis residue volume increase in water

Taking a sample of 100 g pyrolysed resins with a volume of 110 ml, water was added up to a total volume of 400 ml. After 48 hours, no noticeable increase in the volume of the resins could be observed. Evolution of nitrogen gas from the sample was observed and after 48 hours the total volume had been reduced to 370 ml. A likely explanation is that nitrogen was present in the pores of the pyrolysed resins and was replaced by water. The pH of the water after these 48 hours was 9.2, indicating the presence of soluble species in the pyrolysed resins.

5.2. Ignition residue determination

After heating the pyrolysed resins to 550°C for 4 hours in air, 26.5% residue remained. After heating the pyrolysed resins to 900°C for 3 hours in an oxygen atmosphere, 1.36% residue remained. The pyrolysed resins were found to have a sulphate content of 20 ppm and a moisture content of less than 0.05 wt%.

IOP Conf. Series: Materials Science and Engineering **818** (2020) 012008 doi:10.1088/1757-899X/818/1/012008

5.3. Results with anionic resins

A test with anionic resins was done with 50 L resins. Although further testing and sample analysis will be required, the initial results with anionic resins are similar to the results with the cationic resins. A clear volume reduction was observed for the anionic resins. The residue recovered after the test resembled the residue from the cationic resins, with no undesired aggregation of the beads into large chunks.

5.4. Additional tests

During the first quarter of 2020, Montair and Belgoprocess will perform additional tests on the carry over of non-radioactive Co- and Cs-doped anionic and cationic resins, as well as tests with organic absorbents for sludge or liquids.

6. Waste Form

The ashes are a carbon-rich, inert product similar to graphite, free from hydrocarbons and are ideally embedded in cement resulting in an end product ready for disposal. Tests performed on homogeneous cementation of the ashes were successful.

Alternatively, it is also possible to supercompact the drums with the ashes and embed the resulting pucks in cement in a drum. For this latter scenario, the activity levels of the end product need to be watched.

The ashes residue after pyrolysis a VRF of 7 to 8 is measured while test performed with homogeneous cementation an overall VRF from 4 is realised.

7. Conclusions

Montair and Belgoprocess designed, constructed and tested the PRIME pilot installation for treatment of spent resins. This compact installation is transportable in a 20-foot container, which makes the installation mobile and simplifies possible dismantling afterwards. The simple and straightforward process combines the drying of the wet spent resins and pyrolysis in one single reactor without pre-treatment and without intermixing of other waste.

The main features of the installation are a dosing unit with screw feeder which is used for the dosing of the saturated ion exchange resins to the pyrolysis reactor, the pyrolysis reactor itself to treat the resins, an electrically heated thermal oxidizer to treat the gases that are released during drying and pyrolysis, and a wet scrubbing system to scrub the gasses.

The goal of the tests was to investigate the effectiveness of treatment of ion exchange resins, to characterize the end product and to verify that no clogging of individual resin beads occurred during treatment.

The PRIME pilot installation test runs with both cationic and anionic resins were very successful and did not reveal any major issues. Montair and Belgoprocess successfully tested the viability of the installation for the combined drying and pyrolysis process for the resins, the performance of the innovative electrically heated thermal oxidizer and the dosing of the resins using the dosing unit with screw feeder.

The test with cationic resins was performed with 100 L wet resins, which were treated over 10 hours at a maximum temperature of 550°C and under inert nitrogen atmosphere. During pyrolysis, the minimum release of gases was estimated at 81 Nm3 over a period of 4.5 hours, or 18 Nm^3/h. The residue recovered after treatment consisted of dry, ash-coloured beads with almost no agglomeration of resin beads. A volume reduction of 6.1 was obtained over the wet resins. Analysis of the ashes after pyrolysis treatment indicated that the ashes were highly inert, indicating the pyrolysis treatment was successful. A test with 50 L wet anionic resins gave similarly promising results, although further testing and analysis will be required.

Additional tests on the carry over of non-radioactive Co- and Cs-doped anionic and cationic resins, as well as tests with organic absorbents for sludge or liquids are scheduled for the first quarter of 2020.

The results of this pilot plant will be used to realize a mobile commercial installation for the treatment of radioactive spent ion exchange resins from nuclear power plants.

Acknowledgements
This project has received funding from the Euratom research and training programme 2014–2018 under grant agreement No. 755480. This paper reflects only the authors' views, and the European Commission is not responsible for any use that may be made of it.

References

[1] International Atomic Energy Agency 2002 Application of Ion Exchange Processes for the Treatment of Radioactive Waste and Management of Spent Ion Exchangers *Vienna*

[2] Deckers J and Mols L 2007 Thermal Treatment of Historical Radioactive Solid and Liquid Waste into the Cilva Incinerator *ICEM Conference, Bruges, Belgium* pp. 1–6.

[3] Dubois M A, Dozol J F, Nicotra C, Serose J, and Massiani C 1995 Pyrolysis and incineration of cationic and anionic ion-exchange resins - Identification of volatile degradation compounds *J. Anal. Appl. Pyrolysis*, vol. 31, pp. 129–140

[4] Luyckx P, Meeus A, Goetschalckx R, and Hansen J 1999 Pebble Bed Pyrolysis for the Processing of Alpha Contaminated Organic Effluents *WM Symposium*, 1999, Tucson, AZ

IOP Conf. Series: Materials Science and Engineering **818** (2020) 012009 doi:10.1088/1757-899X/818/1/012009

Hot isostatic pressing: thermal treatment trials of inactive and radioactive simulant UK intermediate level waste

L J Gardner[1], S A Walling[1] and N C Hyatt[1]*

[1]Department of Materials Science & Engineering, The University of Sheffield, Mappin Street, Sheffield, S1 3JD.

*n.c.hyatt@sheffield.ac.uk

Abstract. Hot Isostatic Pressing (HIPing) is a batch process thermal treatment technology where wastes are heated and compressed within a sealed stainless steel canister; typically resulting in durable, high density ceramics or glasses with minimal loss of volatile elements, and accountability of active inventories. The University of Sheffield has a small-scale research HIP with capability to process simulant wasteforms containing radioactive materials, to help underpin larger-scale industrial applications of this technology. It was under this remit that a series of trials were undertaken, to produce small simulant radioactive wasteforms incorporating problematic UK waste streams such as Magnox sludges and clinoptilolite ion-exchange material. Each trial was successfully batched, sealed, and HIPed at 1250 °C, resulting in solidified products entirely contained within the steel HIP canisters. The ability to safely produce active wasteforms within the same facility validates the active furnace isolation chamber (AFIC) system. Overall the success of these trials demonstrate the ability of smaller research HIP facilities to build up the scientific and technical case for further implementation of HIP technology as a viable waste treatment option.

1. Introduction

Hot isostatic pressing (HIPing) is a batch thermal treatment technology, which could be utilized to consolidate a wide range of wastes from sludges and spent ion exchange materials, to co-mixed waste streams and for plutonium immobilization, as an alternative to cementation and vitrification technology. The resultant wasteform is typically a durable ceramic or glass-ceramic with hazardous/radioactive elements immobilized within the bulk matrix or bound in tailored crystalline phases. The advantages of HIPing include: batch to batch processing allowing for inventory accountancy, relatively lower processing temperature, high waste loadings, zero off-gas, volume reductions, minimal secondary wastes, and tailorable processing conditions [1]. HIPing for nuclear waste immobilization has largely been limited to inactive surrogates due to the risks associated with handling radioactive compounds and equipment contamination. The Australian Nuclear Science and Technology Organisation (ANSTO) have developed a fully automated HIP process plant for the immobilization of wastes generated from Mo-99 production, due to be completed in 2020 [2]. This will be the first commercial facility available for the processing of radioactive wasteforms via HIPing. Within the UK, the University of Sheffield have the capacity to HIP both inactive and radioactive wasteforms on a prototype scale (~20-50g) using an active furnace isolation chamber (AFIC), which is designed to allow safe use of radioactive materials within a conventional HIP setup, providing a unique opportunity to produce and characterize high waste loaded radioactive simulant wasteforms.

IOP Conf. Series: Materials Science and Engineering **818** (2020) 012009 doi:10.1088/1757-899X/818/1/012009

The EC THERAMIN project (Thermal Treatment for Radioactive Waste Minimization and Hazard Reduction) enabled feasibility demonstration of radiological HIPing with regards to simulant intermediate level wastes. For UK wastes, the scope was limited to magnesium-rich wastes, typically corroded Magnox sludge in addition to clinoptilolite in co-mixed wasteforms. This research program contributed towards the understanding, experience and knowledge gained of radiological HIPing, which could be transferred to large commercial facilities if technology uptake was reached.

2. Methodology

2.1. *Wasteform composition*

Within THERAMIN Work Package 3, three trials were undertaken, utilizing different wastes and targeted chemistries relevant to UK radioactive waste streams. Trial 1, in collaboration with the National Nuclear Laboratory, was a co-mixed product consisting of calcined Magnox sludge (Mg-rich), clinoptilolite, and glass frit, with both CeO_2 as a simulant and U_3O_8. Trial 2 targeted the formation of a magnesium borosilicate glass, using simulant Magnox wastes, boric acid, quartz and U_3O_8; whilst in Trial 3, a simplified composition was targeted, consisting of $Mg(OH)_2$, glass frit and, U_3O_8.

In this study, details are provided for the result of Trial 2 magnesium borosilicate glass wasteforms (MBS-U high and MBS-U low). These glasses were prepared using a combination of magnesium metal and magnesium hydroxide to simulate Magnox sludge – a problematic higher volume intermediate level waste which requires a conditioning route for disposal. Wasteforms were prepared using U_3O_8 to simulate the uranium oxides present in the waste, mostly from corroded mechanically decanned Magnox reactor fuel [3]. Due to the heterogeneous nature of the sludge (which can vary from skip to skip), high and low uranium contents were chosen (42.22 and 6.76 wt. %) to investigate the impact of a varying composition. Any remaining metallic uranium fuel is expected to fully oxidize during the water removal process, which is required during feed preparation. The overall waste loading with respect to magnesium sources and uranium oxide were 59.06 and 40.98 wt. %, respectively. Batching was performed in a controlled laboratory, both wasteforms were pre-calcined in alumina crucibles at 600 °C for 12 hours in a general muffle furnace (air atmosphere) prior to canister packing (discussed in Section 3).

Table 1. Composition of radioactive HIP wasteforms based on a 50g batch

Can No.	Sample Name	$Mg(OH)_2$ (g)	Mg (g)	H_2BO_3 (g)	SiO_2 (g)	U_3O_8 (g)	Waste loading* (wt. %)
17014	MBS-U (high)	5.26	3.16	13.69	6.77	21.11	59.06
17015	MBS-U (low)	16.10	1.01	19.74	9.77	3.38	40.98

*where waste loading refers to $Mg(OH)_2$, Mg and U_3O_8 content

2.2. *Analytical methods*

HIPed canisters were photographed and then sectioned using a Buehler Isomet 1000 precision saw. Powder X-ray diffraction patterns were collected between $5° < 2\theta < 70°$ using a Bruker D2 Phaser diffractometer with a Lynxeye detector and a Ni filtered Cu K_α radiation (1.5418 Å) using a step size of 0.02° 2θ and a count time of 3 s per step. Micrographs were collected using a low vacuum Hitachi TM3030 analytical benchtop SEM with an integrated Bruker EDX system (Quantax 70) at 15 kV and a 7.0 mm working distance on polished (1 μm finish) and carbon coated specimens.

3. Hot isostatic pressing

3.1. *Canister preparation*

Each sample was packed into a straight walled 304 stainless steel HIP canister (*h;* 45 mm, ø; 22 mm) with in-built metal sintered filters. Canister packing was performed using a hydraulic press and die plunger, U_3O_8 samples were prepared in a glovebox to minimize powder distribution/contamination. A hydraulic press was used to maximize the packing density of the wasteform, which also helps to control the canister deformation (i.e. final shape) during the HIP process. Once packed, the canister lid was welded into place using a tungsten inert gas (TIG) welding station. Following this, all canisters were prepared for HIPing by evacuation and bake-out steps. This involved evacuating the canister at room temperature until a vacuum of <8 Pa was achieved after which, the process was repeated at temperature (600 °C) until the initial vacuum recovered. The evacuation tube on the canister was then crimped and welded to produce a hermetically sealed canister ready for processing. These stages are required to remove volatiles/excess water from the wasteform that could affect densification and generate porosity in the final wasteform [4, 5].

3.2. Active furnace isolation chamber

The active furnace isolation chamber (AFIC) is a plug-in component for existing hot isostatic press facilities. The University of Sheffield is uniquely equipped to utilise this containment design, which enables the passively safe processing of uranium-containing wasteforms. The AFIC is engineered to process one canister per HIP cycle, whilst maintaining multiple layers of containment to prevent contamination to the HIP and surrounding work environment in the event of a canister failure, where radioactive powder could be distributed as a result. Other AFIC design features include; a two-way pressure differential, verifiable and changeable filters and seals and, compatibility with glovebox handling operations. Figure 1 showcases the AFIC design, which is protected under patent number WO 2018/009782 A1 [6]. The AFIC was designed for repeated operation without component fatigue beyond typical wear and tear but importantly, the modular design allows for frequent examination and replacement of components without significant cost. Prior to each HIP cycle, the canister is placed in the AFIC and sealed within a controlled laboratory environment before placing into the HIP pressure vessel for processing.

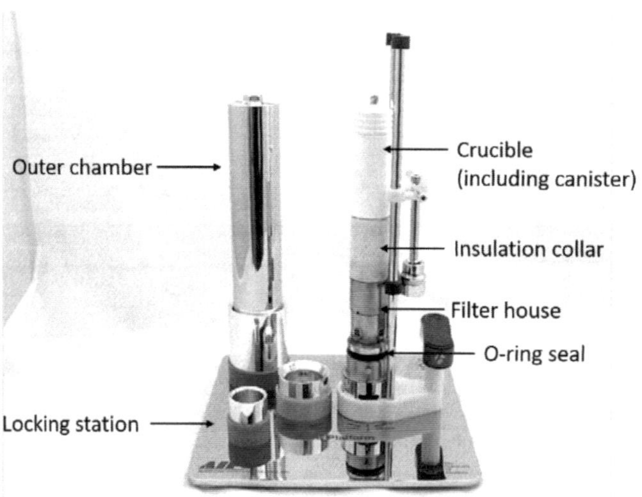

Figure 1. Active furnace isolation chamber (AFIC) components

3.3. HIP process conditions

Hot isostatic pressing was undertaken using an AIP-630H research HIP facility at The University of Sheffield, using argon gas as the pressurizing media. The HIP cycle data are shown in Figure 2 for the magnesium borosilicate glasses, MBS-U high and MBS-U low. The canisters were processed in consecutive stages to protect the welds and control canister deformation to minimise the risk of weld failure. After purging the pressure vessel with argon, the temperature and pressure were ramped to 400 °C and 15 MPa at 5 °C min^{-1} and held for 30 min, followed by a second ramp to 700 °C and 25 MPa at 5 °C min^{-1} and held for 15 min, the final ramp was to 1250 °C and 70 MPa at 7.5 °C min^{-1} and held for 2 hours. The cooling rate was 10 °C min^{-1} and the depressurization rate was -0.6 MPa min^{-1}.

In Figure 2b, it is evident there was an issue maintaining pressure for the MBS-U low wasteform (pressure target was 100 MPa). This was the result of failed compressor packing, which was replaced prior to processing the MBS-U high wasteform, where a controlled pressure ramp and dwell process was achieved. The average pressure during the dwell for the affected wasteform was 77.3 MPa, as such, the target pressure was reduced to 70 MPa for all remaining wasteforms in the work package. Typically, HIP research undertaken at the University of Sheffield targets 100 MPa but the HIP compressor issue limited the maximum controllable pressure to 70 MPa. The difference in pressure is not expected to affect the phase assemblage, which is driven by the temperature and was maintained at target throughout all HIP runs.

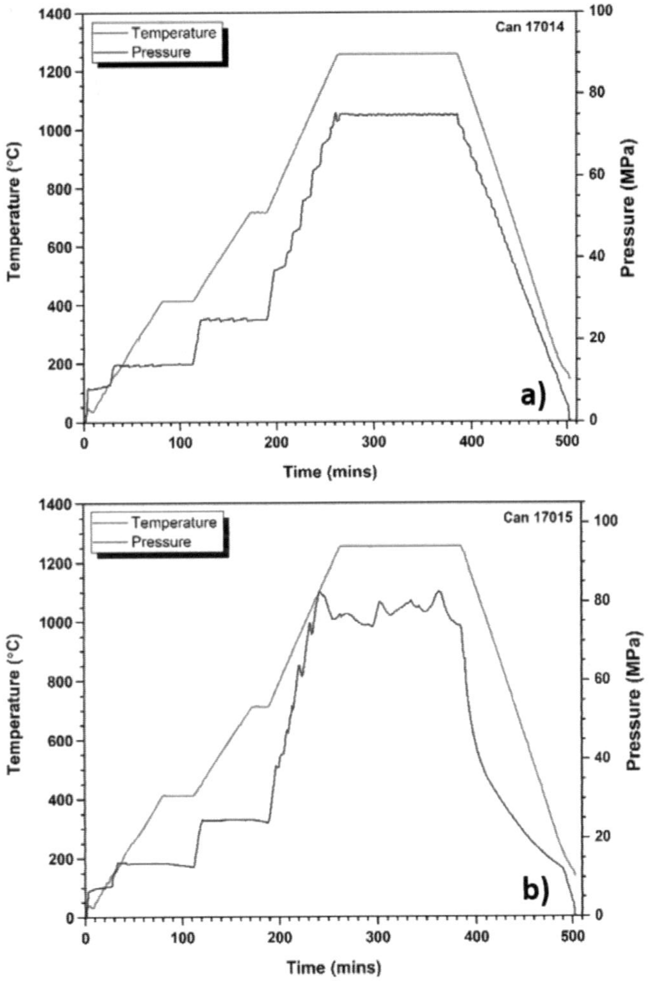

Figure 2. Active HIP process data for canisters 17014 (MBS-U high) and 17015 (MBS-U low)

3.4. Post-HIP observation

The typical appearance of the canisters before and after HIP processing is shown in Figure 3. All canisters in this study were successfully HIPed, this was denoted by the flattened evacuation tube and reduced diameter of the central body and, the canisters remained hermetically sealed (no weld failure). Each canister was sectioned to generate a central slice of the HIPed wasteform (Fig. 3c-d) for characterisation. The sliced sections reveal that both wasteforms formed heterogeneous dense glass/ceramics, with little visible porosity. For MBS-U high (Fig. 3c), the wasteform was largely red/brown in colour with bright white particles distributed throughout, whilst the MBS-U low (Fig. 3d) bulk wasteform was grey in colour with dark grey and bright white particles visible.

Figure 3. Photographs of a HIP canister a) pre-HIP and b) post-HIP and photographs of sectioned HIPed canisters c) MBS-U high and, d) MBS-U low.

3.5. Wasteform characterisation

The phase assemblage for MBS-U high and MBS-U low are reported in Figure 4. For the former, the reflections present were indicative of uranium oxide (likely UO_{2+x}), implying U_3O_8 was reduced by Mg metal during the HIP process. In addition to UO_{2+x}, reflections were assigned to quartz (precursor) and suanite, which is a magnesium borate mineral, $Mg_2B_2O_7$. The lack of intense MgO reflections suggests that both added $Mg(OH)_2$ and Mg metal have reacted, and the targeted magnesium borosilicate glass was formed, with some Mg bound in the suanite phase. For MBS-U low, the phase assemblage was similar, with reflections assigned to UO_{2+x}, quartz and suanite. However, reflections associated with U_3O_8 were also present, indicative of an incomplete uranium reduction compared to MBS-U high. Further data collection is underway to identify the exact uranium oxidation states present in both wasteforms.

The micrographs in Figure 5 highlight the heterogeneous nature of these wasteforms. The large grey particles represent unreacted quartz. This indicates that a more reactive source of silica (e.g. using a simple glass frit) might be more suitable to ensure that the full SiO_2 content was available for glass formation. The bulk matrix is rich in Mg and represents the magnesium based glass and suanite (Fig. 5; Mg). The bright white particles with a varied particle size refer to the UO_{2+x}/U_3O_8 identified by XRD (Fig. 4). Uranium does not appear to have reacted with the MBS components, even in the MBS-U high sample (Fig. 5a) where the uranium loading was high (42.22 wt.%; Table 1). It is postulated that the unreacted quartz resulted in a lower overall vitreous content and minimal uranium incorporation into the glass network. There was no evidence of Mg metal in the processed wasteforms by XRD or SEM/EDX analysis, signifying that the metal was successfully thermally passivated by HIPing. For reactive metal encapsulated in cementitious wasteforms, the volume expansion on oxidation (i.e. Mg to $Mg(OH)_2$), with concomitant H_2 production, could lead to the formation of stress-induced cracking [7]. This oxidation process was completed during HIP processing and the reaction product incorporated into the bulk glass matrix, avoiding any expansion issues.

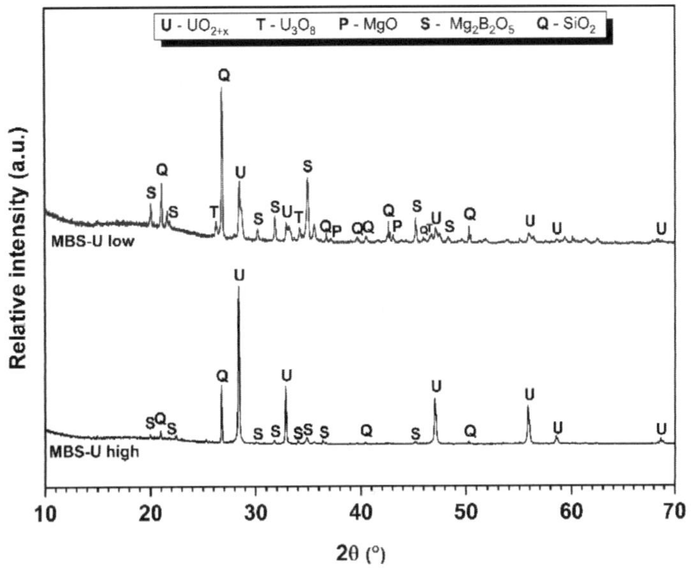

Figure 4. XRD patterns of wasteforms MBS-U high and MBS-U low

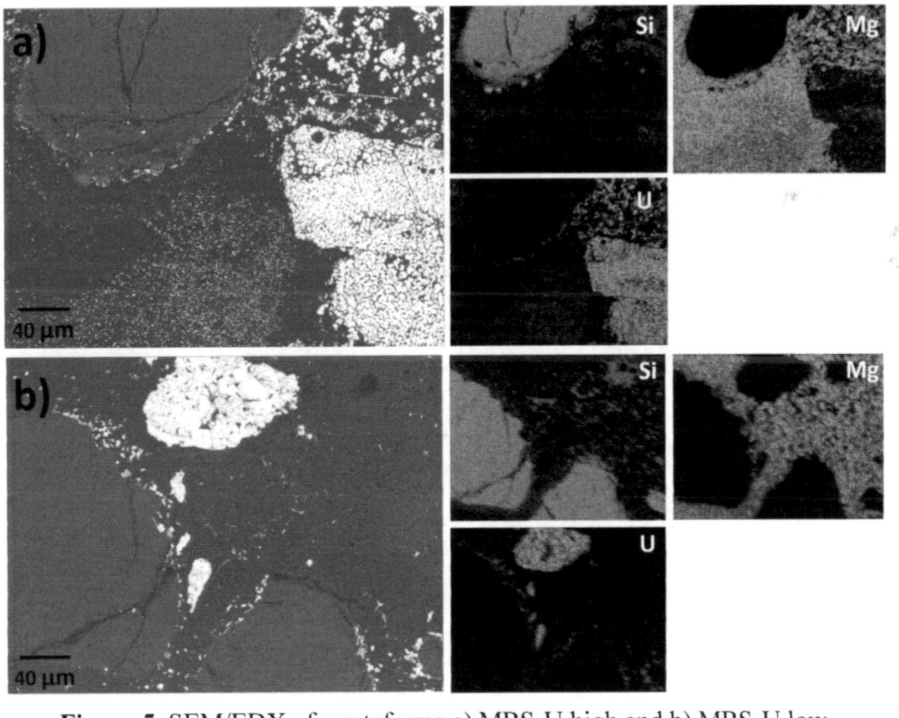

Figure 5. SEM/EDX of wasteforms a) MBS-U high and b) MBS-U low

4. Conclusions

Conceptual wasteforms were prepared and HIPed at The University of Sheffield using radioactive (U_3O_8) waste surrogates for magnesium-rich UK intermediate level wastes. The wasteforms were processed at 1250 °C, ≥ 70 MPa with a 2-hour dwell; all HIPed products exhibited volume reduction and remained hermetically sealed within the stainless steel canister. The processed wasteform was observed to form dense, monolithic glass/ceramics with little visible porosity and good mechanical strength (no fractures formed during sectioning). The magnesium waste ($Mg(OH)_2$, Mg) was fully converted and immobilised in the bulk wasteform, however, U_3O_8 remained undigested (i.e. encapsulated) throughout the encapsulating glass matrix. The wasteform chemistries were not fully optimized during this study but further investigation could improve actinide waste incorporation. HIPing with radioactive waste simulants was successfully demonstrated using the active furnace isolation chamber (AFIC), which permitted the passively safe handling and processing of uranium wasteforms. The experience gained was invaluable to The University of Sheffield and collaborators, leading to greater confidence in the AFIC operational behaviour and contributing towards a technical safety case of hot isostatic pressing as a viable thermal treatment method of nuclear wastes.

5. References

[1] Stewart M W A, Moricca S A, Eddowes T, Zhang Y, Vance E R, Lumpkin G R, Carter M L, Dowson M and James M 2009 *Proc. Int. Conf. Environ. Rem. Radioact. Waste Manage.*, 16523.
[2] Vance E R, Chavara D T and Gregg D J 2017 *MRS Energy Sustain.* **4** E8.
[3] Nuclear Decommissioning Authority 2017 *Waste Stream 2D24: Magnox Cladding and Miscellaneous Solid Waste* (Moor Row, Cumbria: NDA).
[4] Bateman K, Wahlquist D and Malewitz T 2014 *Proc. ASME Int. Mech. Eng. Congr. Expo.* 36935.
[5] Thornber S M, Heath P G, Da Costa G P, Stennett M C and Hyatt N C 2017 *J. Nucl. Mater.* **485** 253.
[6] Moricca S and Persuad R 2018 *World Intellectual Property Organization* WO 2018/009782 A1.
[7] SERCO 2010 *A Survey of Reactive Metal Corrosion Data For Use In The SMOGG Gas Generation Model* (Didcot, Oxfordshire: SERCO).

Acknowledgment

The research leading to these results has received funding from the European Union's Horizon 2020 Euratom research and innovation programme under Grant Agreement No 755480 – THERAMIN. The HIP active upgrade was funded in-part by EPSRC under grant EP/L014041/1 Decommissioning, immobilisation and storage solutions for nuclear waste inventories (DISTINCTIVE). This research utilised the HADES / MIDAS facility at The University of Sheffield established with financial support from EPSRC and BEIS, under grant EP/T011424/1. The authors would like to personally thank Dr. S Moricca and Dr. P. Heath for their continued support with HIP/AFIC operations.

IOP Conf. Series: Materials Science and Engineering **818** (2020) 012010 doi:10.1088/1757-899X/818/1/012010

Hot Isostatically Pressed Zirconolite Wasteforms for Actinide Immobilisation

L R Blackburn[1], L J Gardner[1], S K Sun[1], E R Maddrell[2], M C Stennett[1], C L Corkhill[1], N C Hyatt[1*]

[1]Immobilisation Science Laboratory, Department of Materials Science and Engineering, University of Sheffield, S1 3JD, UK.
[2]National Nuclear Laboratory, Workington, Cumbria, CA14 3YQ, UK.

*n.c.hyatt@sheffield.ac.uk

Abstract. In order to demonstrate the deployment of Hot Isostatic Pressing (HIP) for the immobilisation of Pu stocks and residues, a series of active and inactive zirconolite formulations have been processed and characterised. In this instance, Ce, U, and Th have been applied as chemical surrogates for Pu^{4+}. A range of formulations targeting isovalent Zr^{4+} site substitution (*i.e.* to simulate $CaZr_{1-x}Pu_xTi_2O_7$) have been processed by HIP and characterised by powder X-ray diffraction, and scanning electron microscopy, in order to determine surrogate partitioning between the host zirconolite phase, and accessory phases that may have formed during the HIP process.

1. Introduction

The United Kingdom holds a stockpile of civil separated plutonium that is forecast to exceed 140 teHM (tonnes equivalent heavy metal), subsequent to the completion of domestic reprocessing operations [1]. Immobilisation of Pu within a ceramic matrix such as zirconolite, prototypically $CaZrTi_2O_7$, may be a suitable pathway to disposition. Zirconolite has demonstrated extensive chemical flexibility with regards to feedstock chemistry, and as such is the dominant actinide-bearing phase in the Synroc formulation, developed for the immobilisation of high level waste (HLW) [2]. Pu^{4+} may be accepted into solid solution through Zr^{4+} site, *i.e.* $CaZr_{1-x}Pu_xTi_2O_7$ [3]. The Ca^{2+} site has also been demonstrated to accommodate Pu^{4+} through the addition of a suitable lower valence charge balancing species such as Al^{3+} (*e.g.* $Ca_{0.9}Pu_{0.1}ZrTi_{1.8}Al_{0.2}O_7$) [4]. The preferred final wasteform for Pu would require a robust processing route, insensitive to feed chemistry, with high throughput and waste loading.

Hot Isostatic Pressing (HIP) utilizes the simultaneously application of high temperature and isostatic pressing (typically through the use of an inert medium such as Ar gas) to consolidate and sinter materials to near-theoretical density. The HIP process is insensitive to the chemical properties of the feedstock, and produces minimal secondary waste, and is therefore a good candidate thermal treatment for the immobilisation of Pu stocks and residues [5]. In the current work, we report laboratory scale trials for zirconolite wasteforms processed by HIP. In this instance, CeO_2, UO_2 and ThO_2 were employed as chemical surrogates for PuO_2.

IOP Conf. Series: Materials Science and Engineering **818** (2020) 012010 doi:10.1088/1757-899X/818/1/012010

2. Materials synthesis and characterisation

2.1. Materials synthesis
Zirconolites were batched from oxide precursors ($CaTiO_3$, ZrO_2, TiO_2, CeO_2, ThO_2, UO_2) targeting nominal stoichiometry $CaZr_{0.8}Ce_{0.2}Ti_2O_7$, $CaZr_{0.8}U_{0.2}Ti_2O_7$ and $CaZr_{0.8}Th_{0.2}Ti_2O_7$. Precursor materials were optimised for HIP according to our previous investigations considering zirconolite glass-ceramic wasteforms [6]. Materials were batched and homogenised by high intensity ball milling (400 rpm, ~ 40 min), with the addition of isopropanol as a milling agent. Powder cakes were dried and passed through a sieve to break up agglomerates. Materials were then calcined at 600 °C in air for 12 h (Ar gas was used in the case of $CaZr_{0.8}U_{0.2}Ti_2O_7$ in order to maintain U^{4+}), before being pressed into the walls of a HIP can (316 stainless steel), under ~ 1 t uniaxial pressure. The cans were then evacuated at 300 °C for a further 12 h until approximately 50 mTorr pressure was achieved. The HIP cans were then hermetically sealed and placed into a AIP 630-H HIP unit and consolidated at 1300 °C and 100 MPa for 4 h.

2.2. Materials characterisation
A portion of the bulk ceramic was cut free from stainless steel containment and prepared for microchemical analysis using a Hitachi TM3030 scanning electron microscopy (SEM), fitted with a Bruker Quantax Energy Dispersive X-ray Spectrometer (EDS) for compositional analysis. The sample was mounted in cold setting resin and polished to a 1 μm optical finish using incremental grades of diamond suspension paste. SEM analysis was undertaken with a 15 kV accelerating voltage, and a working distance of ~ 8mm. A specimen was also mechanically ground prior to powder X-ray diffraction analysis using a Bruker D2 Phaser, using Cu Kα radiation ($\lambda = 1.5418$ Å). This was fitted with a Lynxeye position sensitive detector. Data were acquired in the range $5^\circ \leq 2\theta \leq 80^\circ$, with step size 0.02°.

3. Results

3.1. $CaZr_{0.8}Ce_{0.2}Ti_2O_7$
The HIPing process was successful in all instances, and an example of a canister pre and post-processing is displayed in Figure 1. Powder X-ray diffraction data for $CaZr_{0.8}Ce_{0.2}Ti_2O_7$ are displayed in Figure 2. Refinement of the powder diffraction profile determined that zirconolite-2M was formed as the major crystalline phase – these reflections are indexed by relevant (hkl) values. Reflections consistent with the zirconolite-4M polytype phase were also observed, noted by (002) and (008) supercell reflections at $2\theta = 7.8^\circ$ and $2\theta = 31.1^\circ$, respectively. The formation of the zirconolite-4M polytype is consistent with previous observations, with the incorporation of tetravalent species within the Zr^{4+} site [7,8]. A considerable inventory of perovskite (parent structure $CaTiO_3$) was also stabilised; the formation of accessory perovskite phases is commonly attributed to the partial reduction of Ce^{4+} to Ce^{3+} [7,9]. XANES data (not shown) confirmed that a considerable fraction of Ce^{4+} was reduced to the Ce^{3+} species. EDS spotmap analysis of the perovskite phase, in this instance, confirmed the partial retention of Ce. Microscopic analyses (Figure 3) were consistent with powder XRD data. Zirconolite-4M and perovskite were clearly identified, distinguished by variation in backscattered electron contrast; these phases were consistently enriched in Ce.

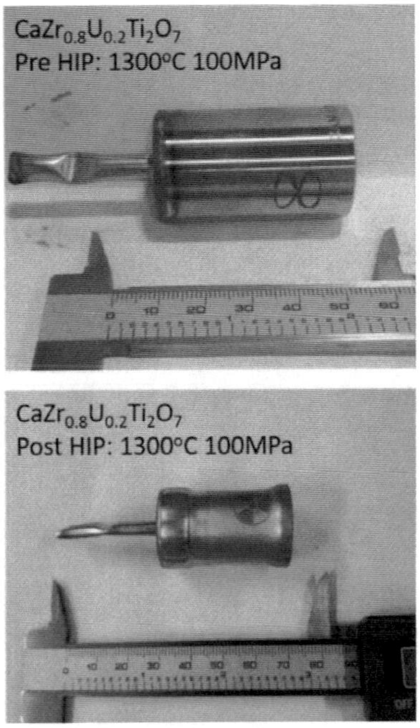

Figure 1. Photograph of HIP canister pre and post-processing at 1300 °C and 100 MPa for 4 h.

3.2. $CaZr_{0.8}U_{0.2}Ti_2O_7$

When targeting nominal stoichiometry $CaZr_{0.8}U_{0.2}Ti_2O_7$, zirconolite-2M was formed as the dominant U-bearing phase, as determined by powder X-ray diffraction (Figure 2). Weak reflections indicative of a $(Zr,U)O_2$ solid solution were also identified. An ancillary perovskite phase was evidenced by powder XRD data (diagnostic reflection $2\theta = 33.1°$) however, this was not clearly identified by SEM analysis. The zirconolite-4M polytype was not observed to form, contrary to experimental design and previous observations by Vance *et al.* [8]. Furthermore, UO_2 did not appear to be fully accommodated within the zirconolite phase, inferred by reflections consistent with the cubic UO_2 structure present in XRD analyses; SEM observations confirmed that free UO_2 was distributed throughout the microstructure (Figure 4). The $(Zr,U)O_2$ solid solution appeared to form at the rim of undigested UO_2 relics within the microstructure.

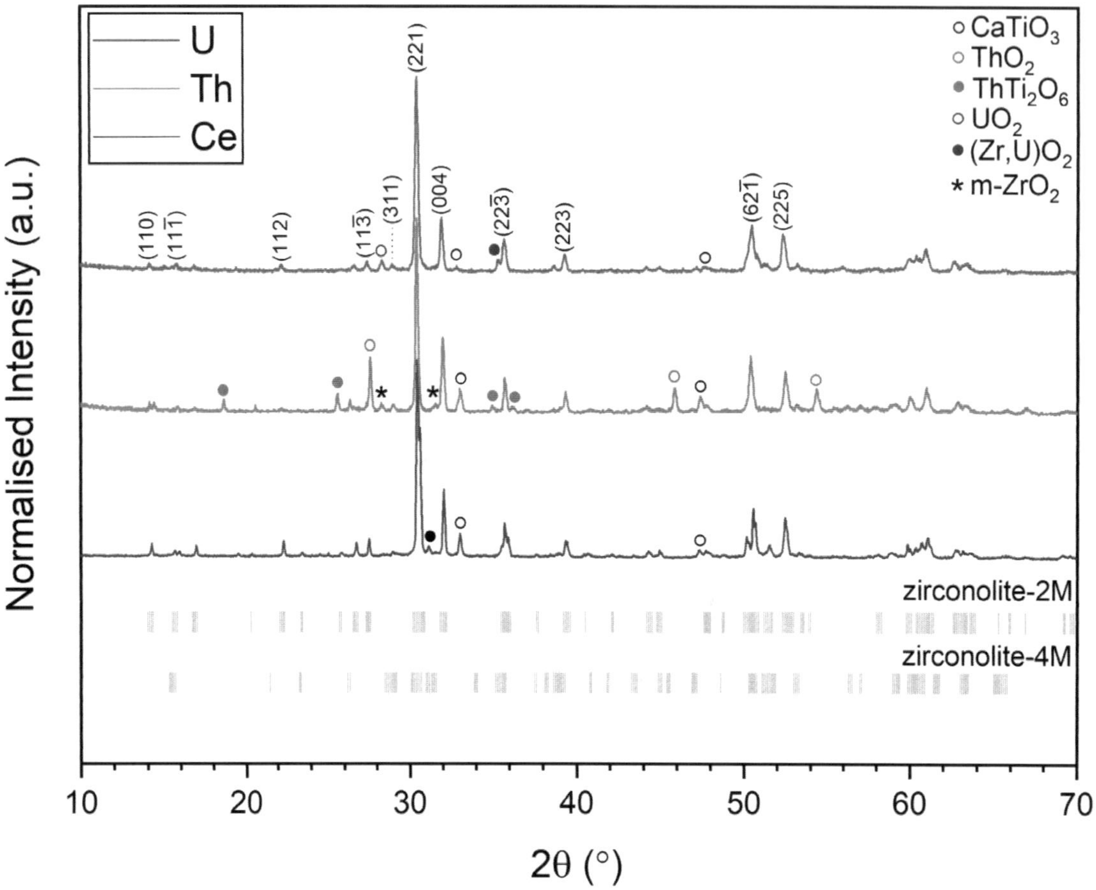

Figure 2. Powder X-ray diffraction data for $CaZr_{0.8}Ce_{0.2}Ti_2O_7$, $CaZr_{0.8}U_{0.2}Ti_2O_7$ and $CaZr_{0.8}Th_{0.2}Ti_2O_7$ processed by HIP at maximum temperature 1300 °C, at 100 MPa for 4 h. Zirconolite-2M reflections are labelled by (hkl) values; zirconolite-4M reflections are labelled by closed circles (•).

3.3. $CaZr_{0.8}Th_{0.2}Ti_2O_7$

Phase determination for $CaZr_{0.8}Th_{0.2}Ti_2O_7$ from powder XRD data are displayed in Figure 2. Diagnostic reflections consistent with zirconolite-2M were identified; these are labelled by appropriate (hkl) values. Intense (111) and (022) reflections, coherent with the fluorite structured ThO_2 phase, were clearly evidenced at $2\theta = 27.5^{\circ}$ and $2\theta = 45.7^{\circ}$, showing that ThO_2 was not fully incorporated within the zirconolite phase. This was confirmed by SEM analyses – see Figure 5. An ancillary perovskite phase was also clearly evidenced by powder XRD and SEM measurements. The zirconolite-4M polytype did not appear to form; a secondary thorutite phase, with ideal stoichiometry $ThTi_2O_6$, was stabilized. This phase was distributed around the rim of undigested ThO_2 within the microstructure as is illustrated in Figure 5.

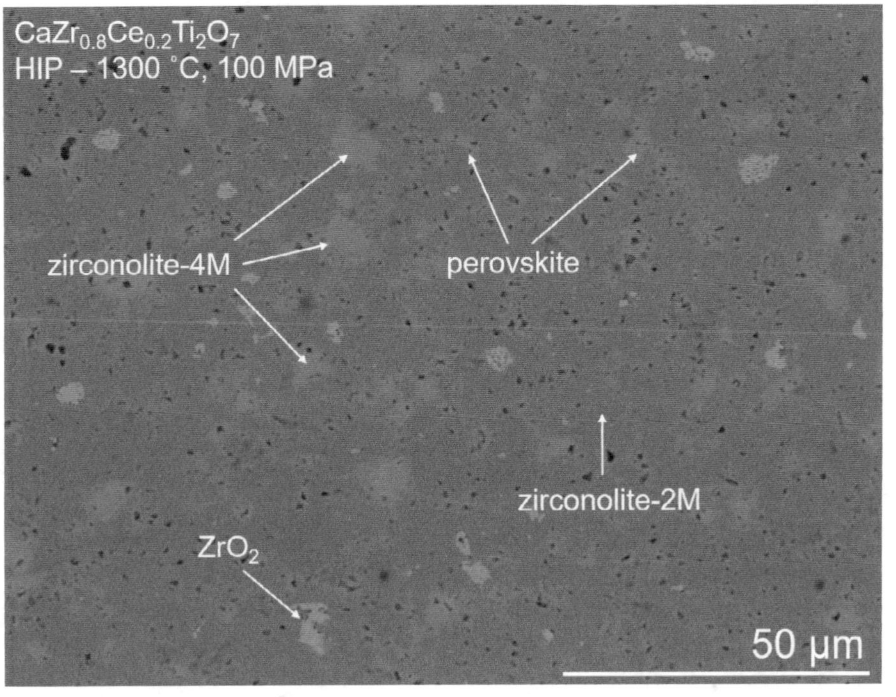

Figure 3. Representative backscattered electron micrograph of the $CaZr_{0.8}Ce_{0.2}Ti_2O_7$ microstructure formed by HIP at 1300 °C at 100 MPa

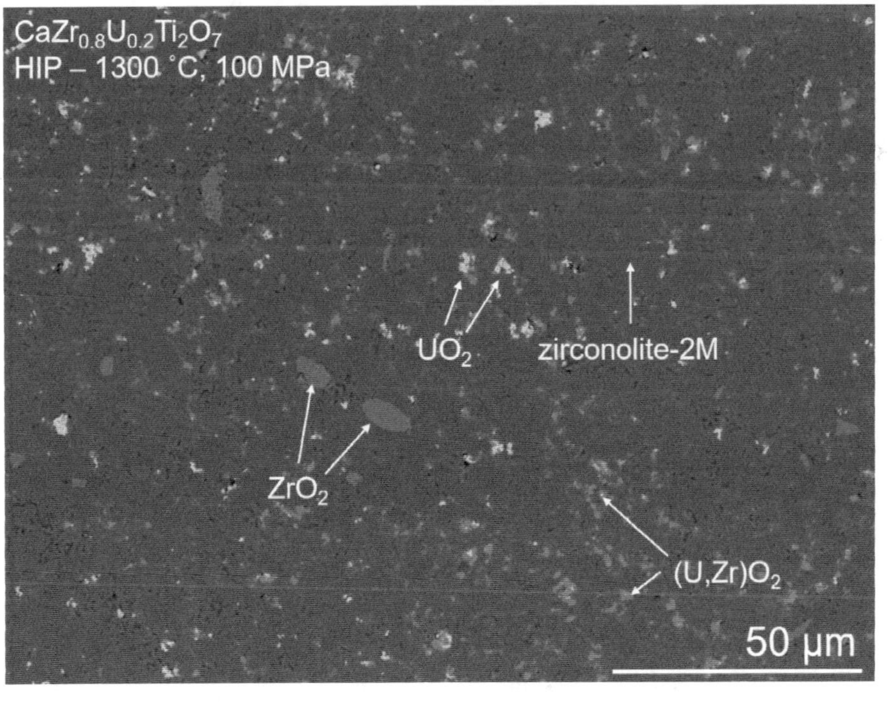

Figure 4. Representative backscattered electron micrograph of the $CaZr_{0.8}U_{0.2}Ti_2O_7$ microstructure formed by HIP at 1300 °C at 100 MPa

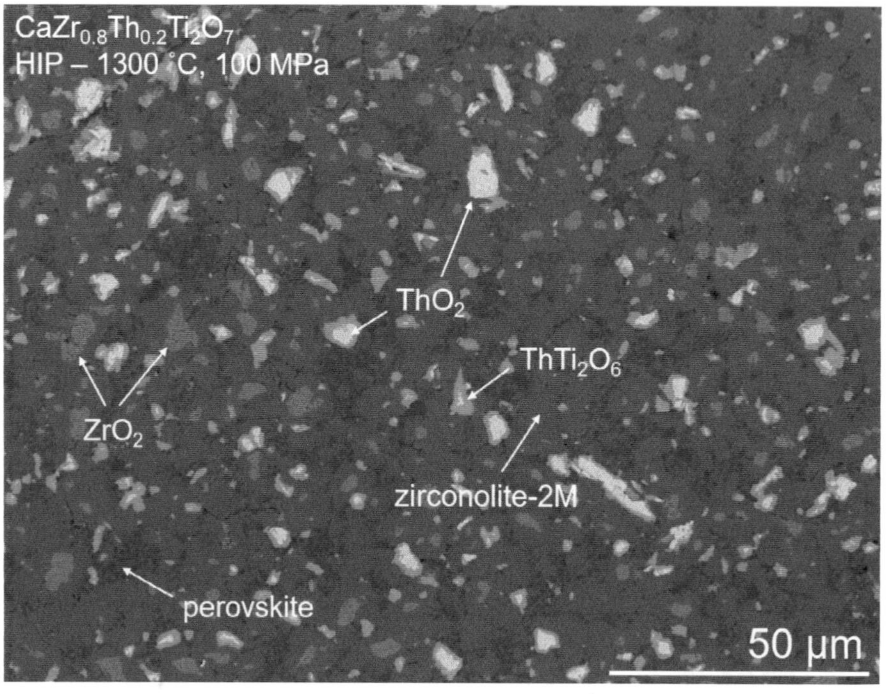

Figure 5. Representative backscattered electron micrograph of the $CaZr_{0.8}Th_{0.2}Ti_2O_7$ microstructure formed by HIP at 1300 °C at 100 MPa

4. Conclusions

Zirconolite wasteforms incorporated with CeO_2, UO_2 and ThO_2, chemical surrogates for PuO_2, have been processed by laboratory scale HIP. The processing parameters utilised a maximum temperature of 1300 °C, under an Ar pressure of 100 MPa; these conditions were held for a dwell time of 4 h. A mixture of zirconolite-2M and zirconolite-4M polytypes were observed when targeting nominal formulation $CaZr_{0.8}Ce_{0.2}Ti_2O_7$; a notable fraction of Ce-perovskite was also stabilised. The formation of perovskite is undesirable, due to inferior chemical durability with respect to the target zirconolite phase [10]. The crystallisation of zirconolite-4M was not observed in the case of $CaZr_{0.8}U_{0.2}Ti_2O_7$ or $CaZr_{0.8}Th_{0.2}Ti_2O_7$, due to the incomplete accommodation of UO_2 and ThO_2 within the zirconolite phase. The incorporation of UO_2 and ThO_2 within zirconolite stabilized secondary $(Zr,U)O_2$ and $ThTi_2O_6$ phases, respectively. The use of multiple surrogates may allow the imply the possible localization of Pu within a similar system. The reducing conditions imposed by the HIP environment may promote the formation of Pu^{3+}, and subsequent partitioning within the ancillary perovskite phase, decreasing the wasteform stability under hydrothermal conditions. These data demonstrate the need for careful phase design, in order to suppress the formation of deleterious accessory phases.

5. References

[1] Nuclear Decommissioning Authority (NDA), *Progress on Plutonium Consolidation, Consolidation, Storage and Disposition*, 2019.

[2] Ringwood A E, Oversby V M, Kesson SE, Sinclair W, Ware N, Hibberson W and Major A 2005 *Nucl. Waste Manag.* **2** 287.

[3] Begg B D, Day R A and Brownscombe A 2001 *Mat. Res. Soc. Symp. Proc.* **663** 1.

[4] Begg B, Vance E. and Conradson S. 1998 *J Alloys Compd.* **271–273**, 221.

[5] Maddrell E. 2012 *Chem. Eng. Res. Des.* **91** 735.

[6] Thornber S, Maddrell E, Stennett M C and Hyatt N C, 2016, in *Proceedings of the Waste Management Symposium 2016*, Phoenix, AZ, USA, 6 – 10 March 2016, 16581.

IOP Conf. Series: Materials Science and Engineering **818** (2020) 012010 doi:10.1088/1757-899X/818/1/012010

[7] Clark B M, Sundaram S K and Misture S T 2017 *Sci. Rep.* **7** 2.
[8] Vance E R, Lumpkin G R, Carter M L, Cassidy D J, Ball C J, Day R A and Begg B, 2002 *J. Am. Ceram. Soc.* **85** 1853.
[9] Begg B D and Vance E R, 1997 *Mat. Res. Soc. Symp. Proc.* **465** 333.
[10] McGlinn P J, Hart K P, Loi E H and Vance E R, 1995 *Mat. Res. Soc. Symp. Proc.* **353** 847.

Acknowledgement
We are grateful for financial support from the Nuclear Decommissioning Authority and EPSRC under grant numbers EP/201019X/1, EP/N017870/1 and EP/R511754/1. This research utilised the HADES/MIDAS facility at the University of Sheffield established with financial support from EPSRC and BEIS, under grant EP/T011424/1.

Advanced Gas-cooled Reactor SIMFuel Fabricated by Hot Isostatic Pressing: a Feasibility Investigation

H Ding,[1] L J Gardner,[1] R Mohun,[1] C L Corkhill,[1] M C Stennett,[1] N C Hyatt[1]*

[1]Department of Materials Science & Engineering, The University of Sheffield, Mappin Street, Sheffield, S1 3JD.

*n.c.hyatt@sheffield.ac.uk

Abstract. The manufacture of a simulant UK Advanced Gas Cooled Reactor (AGR) spent nuclear fuel (SIMFuel) was achieved by Hot Isostatic Pressing (HIP). Characterisation of HIP AGR SIMFuels, tailored to burn ups of 25 GWd/t U and 43 GWd/t U (after 100 years cooling) demonstrated fission product partitioning, phase assemblage, microstructure and porosity in good agreement with spent nuclear fuels and SIMFuels, and AGR fuels in particular. A pivotal advantage of the application of the HIP manufacturing method is the retention of volatile fission products within the resultant SIMFuel as the result of using a hermetically-sealed container. This new approach to SIMFuel manufacture should enable the production of more accurate spent nuclear fuel surrogates to support research on spent fuel management, recycle, and disposal, and the thermal treatment of fuel residues and debris.

1. Introduction

The safe management and disposal of spent nuclear fuel is a critical challenge for nation states exploiting civil nuclear energy and the current global inventory of spent nuclear fuels exceeds 300,000 metric tonnes of heavy metal [1,2]. Research involving spent nuclear fuels is challenging and expensive, due to their extremely high dose rate, which requires specialist remote handling facilities. Consequently, there is a need to develop a mechanistic understanding of spent nuclear fuel behavior under conditions relevant to geological disposal, using low activity analogues, which accurately capture the chemistry, phase assemblage and microstructure [3-7]. Such simulant spent nuclear fuels, colloquially referred to as SIMFuels, are also of importance for development of reprocessing flow sheets and thermal treatment processes applied to damaged and degraded nuclear fuels, fuel debris, and fuel residues.

A challenge posed by SIMFuels developed to date is the absence of incorporated volatile fission products, such as Cs, I, Te, and Mo. Cs and I fission products are of critical importance for the early release fraction of spent nuclear fuels in the disposal environment, when infiltrating groundwater encounters a corroded or breached fuel assembly. The typical fabrication process for SIMFuels involves the high temperature (>1600°C) solid state reaction of UO_2 and stable fission product oxides (or alternate reagents), and sintering for several hours, under reducing atmosphere, to give a ceramic body of appropriate porosity. Under such conditions, volatile fission products are completely or partially lost from the phase assemblage [3-7].

Recently, Wangle *et al.* reported the fabrication of SIMFuel material incorporating Cs and I stable fission products, using spark plasma sintering (SPS) [8]. In this approach, an intimate mixture of UO_2 and CsI was subject to reactive sintering in a graphite die, at 1000°C and 10 min under a holding pressure of 17.7 MPa. The retention of CsI as a separate phase was demonstrated by SEM/EDX and XRD investigation. Nevertheless, some evaporation of CsI from the SPS set up was observed by the authors and the retained CsI was determined to be lower than the initial content. However, this approach has not yet been applied to more complex and realistic SIMFuel chemistries, with a more complete fission product inventory.

We considered that retention of the volatile fission product inventory in SIMFuel materials could be achieved under conditions of Hot Isostatic Pressing (HIP), in which reactive sintering is typically achieved in a hermetically-sealed stainless steel can, under high temperature and pressure (for fabrication of radioactive wasteforms: 1250°C and 200 MPa Ar) [9]. Our development of this technology for the thermal treatment of radioactive wastes has exploited this advantage to contain volatile radionuclide constituents within the product wasteform. Accordingly, we have undertaken a feasibility study of SIMFuel fabrication, incorporating volatile fission products, by HIP. We focused on the fabrication of SIMFuels for UK Advanced Gas Cooled Reactors (AGRs) for which there exists comparative data for counterparts produced by conventional reactive sintering under reducing atmosphere [10]. AGRs are graphite moderated and CO_2 cooled reactors, which utilize UO_2 fuel, enriched to 2.5-3.5% ^{235}U clad in stainless steel [11]. Under conventional SIMFuel fabrication conditions, complete volatilization of Cs and Te was observed, and the retained content of Mo, Ru, Rh and Pd was reported to be lower than expected, possibly due to volatilization [10]. For the purpose of this feasibility study, a simplified SIMFuel chemistry was used, with the focus on understanding the retention of Cs, Te and Mo as volatile fission products of interest.

2. Experimental

2.1. Preparation of SIMFuel feasibility samples

This investigation considered fabrication of AGR SIMFuel with a burn up of 25 GWd/t U and 43 GWd/t U, after 100 years of cooling. The estimated stable fission product inventory of such spent nuclear fuels was previously determined using the FISPIN 10.0.1 and reported by Hiezl *et al.* [10]. Based on the published estimated inventory, a simplified SIMFuel composition was established, Table 1, omitting PdO and Rh_2O_3, which are in any case known to form alloy inclusions within SIMFuel and spent nuclear fuels. The depleted UO_2 (>99.9%) powder used in this experiment was purchased from Absco limited (Haverhill, Suffolk, UK) and stable fission product surrogates were purchased from either Alfa Aesar (Heysham, Lancashire, UK) or Sigma Aldrich (Gillingham, Dorset, UK).

Table 1. Target compositions of two SIMFuels after 100 years cooling

SIMFuels composition	25 GWd/t U		43 GWd/t U	
	at.%	wt%	at.%	wt%
UO_2	97.587	98.193	95.870	97.586
Nd_2O_3	0.483	0.605	0. 827	0.364
ZrO_2	0.369	0.169	0.602	0.279
MoO_3	0.334	0.179	0.566	0.307
RuO_2	0.257	0.127	0.454	0.228
BaO	0.244	0.179	0.435	0.323
CeO_2	0.193	0.124	0.329	0.213
La_2O_3	0.096	0.117	0.160	0.196
SrO	0.032	0.012	0.050	0.019
Y_2O_3	0.041	0.034	0.064	0.054
Cs_2O	0.191	0.231	0.309	0.379
TeO_2	0.044	0.026	0.079	0.047

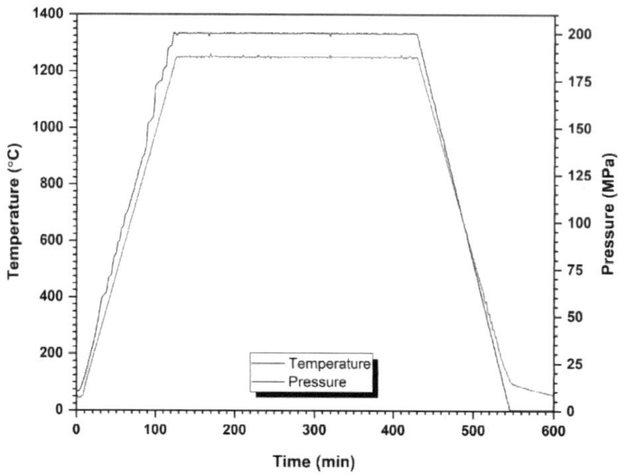

Figure 1. Heating and pressure profile of HIP cycle.

The fission product surrogates were stable oxides except for the use of $BaCO_3$ and Cs_2CO_3. A master batch of UO_2 stable fission product oxides / carbonates was produced according to the composition in Table 1 and planetary ball-milled for 15 minutes at rate of 500 rpm. The required quantities of blended stable fission product and remaining UO_2 reagent powders were then planetary milled for 30 minutes at 500 rpm. The homogenised batches were packed into a stainless steel tube (diameter: inner – 3mm, outer – 5 mm) and subject to bake out under 5 Pa vacuum at 500°C for 5 hours. This temperature was selected to be below the decomposition temperature of $CsCO_3$ to avoid volatilization of Cs (610 °C). The stainless steel tube (Type 316L) was then crimped and welded closed and subject to hot isostatic pressing. The HIP cycle is shown in Figure 1 and incorporated a 5-hour dwell at 1250 °C/200 MPa, with a 10 °C min^{-1} heating and cooling rate. The ramping pressure was controlled manually and released automatically to match the temperature change as shown in Figure 1. The total volume of SIMFuel produced for each material was approximately 100 mm^3.

2.2. Characterisation techniques
Following the HIP cycle, the stainless steel tubes were sectioned to access the HIPed SIMFuel material for characterisation. Powder material was carefully removed from the stainless steel tube, crushed and analysed with a Bruker D2 Phaser diffractometer, using Ni filtered Cu Kα radiation (1.5418 Å). All X-ray diffraction patterns (XRD) were collected in the range $10° \leq 2\theta \leq 80°$ with a step size of 0.02°. Unit cell parameters were determined by Rietveld profile analysis using a LaB_6 internal standard. Tube cross-sections were mounted in cold setting resin, polished and characterized by Scanning Electron Microscopy with Energy Dispersive X-ray analysis (SEM/EDX). Backscattered electron images were acquired using a Hitachi TM3030 SEM with an accelerating voltage of 15 kV, to investigate the microstructure. Images with same magnification were also used to estimate the porosity of the SIMFuel based on the threshold of grey level obtained from ImageJ software. Raman spectroscopy was used to identify changes in the local structure of the UO_2 matrix. Raman spectra were gathered in the range 200-800 cm^{-1} using a Horiba XploRA Plus Raman microscope with a 523 nm laser and analysed by deconvolution of the band features.

3. Results and Discussion

3.1. Microstructure and porosity
The microstructure of the HIP SIMFuel samples were observed by SEM to exhibit a grain size of ca. 1 μm as shown in Figure 2. The porosity of the SIMFuel determined by image analysis, using ImageJ software, are reported in Table 2, together with that determined for real AGR spent nuclear fuel. It is

apparent that the porosity of the HIP SIMFuels are in good agreement with real AGR spent nuclear fuel and much smaller that the porosity reported for conventionally sintered AGR SIMFuels, typically ca. 20% [12, 13]. The smaller standard deviation in the porosity of the HIP AGR SIMFuel may be due to the homogenised fabrication method, whereas temperature and burnup (and hence composition) gradients exist in the real spent nuclear fuel. It was found that porosity increases for the high burn up composition, which could be attributed to the higher volatile fission product content (because the volatile fission products formed gas bubbles as HIP tube was hermetically sealed).

Table 2. Statistics of area parentage and lattice parameter of HIP AGR SIMFuels.

	25 GWd/t U	43 GWd/t U	Ref
Area Porosity %	2.17 ± 0.32	2.85 ± 0.33	2.19 ± 0.84 [12]
Unit cell parameter a (Å)	5.4721 ± 0.0001	5.4753 ± 0.0001	5.4705 ± 0.0002 [13]

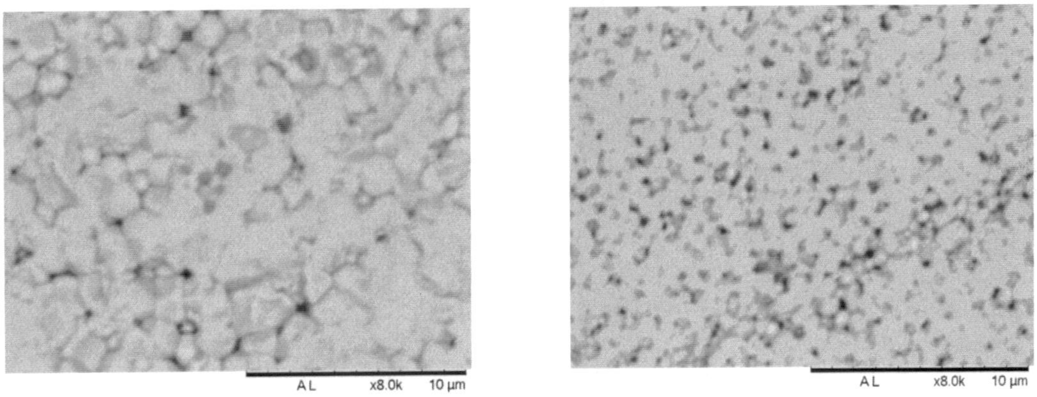

Figure 2. Representative SEM images of microstructure and porosity in 25 GWd/t U (left) and 43 GWd/t U (right) HIP AGR SIMFuels

3.2. UO_2 matrix

The XRD patterns of the 25 GWd/t U and 43 GWd/t U HIP SIMFuel samples exhibited the reflections characteristic of UO_2 (PDF 03-065-028) without additional reflections, as shown in Figure 3. The concentration of additional phases formed by stable fission products was evidently below the level of detection and resolution of the instrument.

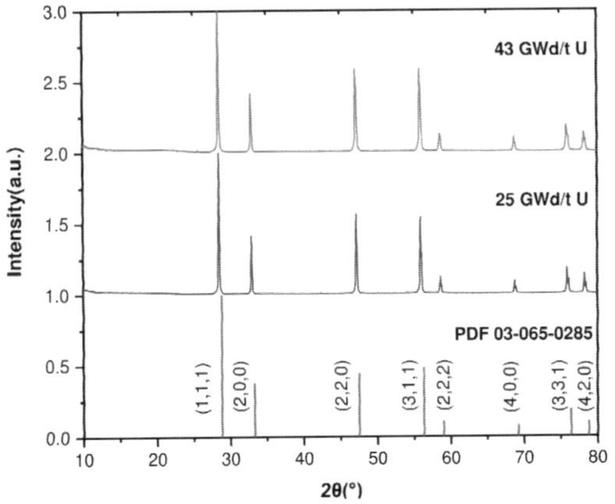

Figure 3. XRD patterns of HIP AGR SIMFuel samples with reference UO_2 pattern.

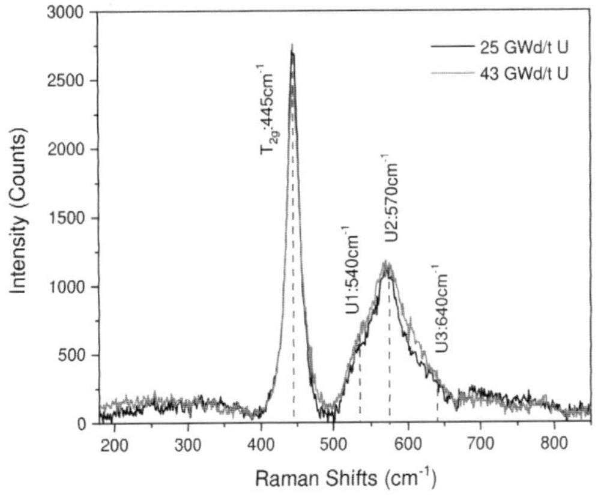

Figure 4. Raman spectra of HIP AGR SIMFuel samples and bands identification.

The refined unitcell parameter (Table 2) showed a slight increase (0.0320 ± 0.0001 Å) from low-doped sample to high-doped sample. It matches well with the typical lattice parameters variations for the intermediate pellet region in real spent nuclear fuel reported in previous study [14,15]. The change is probably due to the fission products (Nd, La, Ce, Zr Y) dissolved in the UO_2 matrix enlarging the unit cell with a larger average ionic radius. The average ionic radius of dissolved fission products was calculated to be 1.040 Å and 1.042 Å for the 25 GWd/t U and 43 GWd/t U HIP AGR SIMFuels, respectively, compared to 1.000 Å for U^{4+} (all in eight-fold co-ordination). It was found that the change of average ionic radius follows the change of lattice parameter but does not yield a strict linear relationship. The potential cause is the different solubility of fission products as some (e.g. Zr, Ce) were found to be partitioned into oxide precipitates (see below).

Raman spectroscopy analysis, Figure 4, provided further evidence of stable fission product incorporation within the UO_2 matrix, consistent with XRD data. As shown in Figure 4, the sharp T_{2g} band, at 445 cm^{-1}, was observed and assigned to the U-O stretching mode within a symmetric UO_8 cube of the UO_2 fluorite structure [16-18]. The broad feature in the range 500 - 650 cm^{-1} comprises three bands which are sensitive to defect induced distortion of the UO_2 fluorite lattice, assigned as: U_1 at 540 cm^{-1}, due to oxygen defects; U_2 at 570 cm^{-1}, longitudinal optical mode; and U_3 at 640 cm^{-1} due to oxygen interstitials. The relative intensity of the broad feature comprising the U_1, U_2, and U_3 bands increases, relative to the intensity of the T_{2g} band, with increasing burn up. This is consistent with an increased incorporation of fission product species within the UO_2 matrix and hence an increased concentration of lattice defects. The oxygen defect distortion might be induced by fission products dissolved in the UO_2 matrix with an average valence state below U^{4+}, which is consistent with previous Raman analysis of SIMFuel [19].

3.3. Analysis of fission product oxide and metallic precipitates

Fission product oxide and metallic precipitates were formed and observed to be uniformly distributed in the UO_2 matrix. In spent nuclear fuel and SIMFuels, the formation of a perovskite oxide "grey phase" incorporating Zr, Ba, Sr, Mo and Ce, at the boundaries of grains and pores, has been reported [4-7, 10]. Figure 5 shows such a typical grey phase precipitate, a few µm in size. The EDX spectrum of this phase, Figure 5, confirmed a composition consistent with that expected for the grey phase, $(Ba,Sr)(Zr,U)O_3$ as reported in previous study[10]. Additionally, trace Y, Ce, and Mo were detected within some grey phase oxide precipitates by spot EDX analysis, but the major compositions of these grey phases were similar.

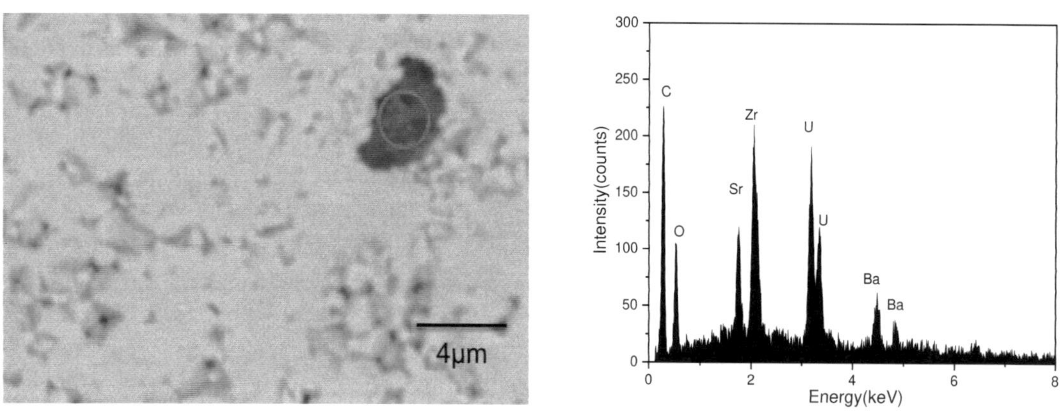

Figure 5. Back scattered electron image and EDX spectrum of typical grey phase oxide precipitate in HIP AGR SIMFuel; in this example 43 GWd/t U.

Metallic precipitates observed in the HIP AGR SIMFuels presented variable morphology and size. EDX analysis of the metallic precipitates revealed that the primary components were typically Mo and Ru.

The elemental compositions were variable but the content of Ru was generally lower than that of Mo, which agrees well with the previous study [4, 10]. Figure 6 shows a typical metallic precipitate with about 64 at. % Mo and 8.7 at. % Ru.

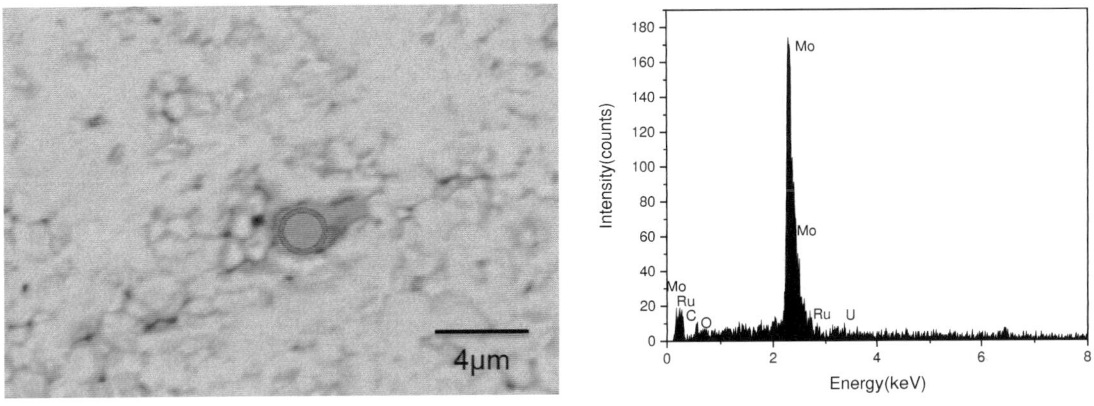

Figure 6. Back scattered electron image and EDX spectrum of typical meallic precipitate in HIP AGR SIMFuel; in this example 43 GWd/t U.

3.4. Analysis of cesium retention

Cs was found to be enriched in the gap between the pellet and stainless steel tube, as shown in the Figure 7, which is consistent with the behaviour observed in AGR spent fuels as well as Boiling and Pressurised Water Reactor (BWR and PWR) spent fuels [1, 20-24]. In the case of AGR fuels, it is hypothesised that Cs is transported as volatile Cs_2MoO_4, with the formation of Cs_2MoO_4, CsI or Cs_2Te at the fuel pellet rim [20]. In the present HIP AGR SIMFuels, Cs was found to be associated with Cl, implying the formation of CsCl at the interface between the SIMFuel and stainless steel surface. The formation of CsCl has not been previously reported in post irradiation examination of spent nuclear fuels or characterisation of SIMFuels. However, Cl rich deposits were reported on the internal surface of the Zircaloy cladding, some of which were associated with a high concentration of Cs, although the contribution of other elements to the wavelength dispersive X-ray fluorescence spectra made conclusive identification of the Cl bearing phase difficult [21].

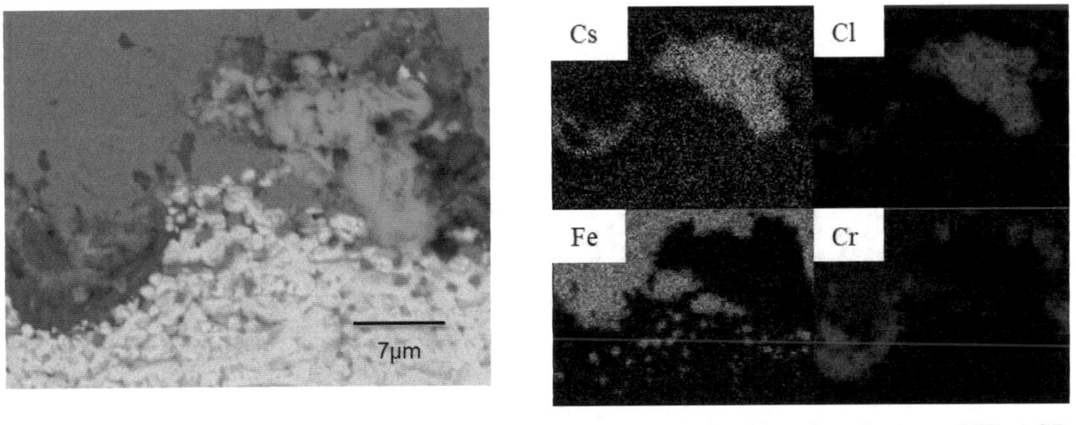

Figure 7. Back scattered electron image and elemental mapping of interface between HIP AGR SIMFuel 43 GWd/t U and stainless steel tube.

IOP Conf. Series: Materials Science and Engineering **818** (2020) 012011 doi:10.1088/1757-899X/818/1/012011

The origin of the chlorine in the current observations is uncertain, but could conceivably arise from contamination of reagents, equipment, processes or the stainless steel tubing. The Cl content of UO_2 nuclear fuels is typically trace, for example < 5ppm for CANDU fuel and ^{36}Cl (produced by neutron activation of ^{35}Cl) is of concern for the instant release fraction of spent nuclear fuel [25]. Thus, it is plausible that CsCl is a candidate for rim-enriched Cs species in AGR fuels. However, it is recognised that these species remain poorly characterised in real AGR spent fuels, though it is known that partitioning of Cs into a separate layer on the rim surface does occur.

4. Conclusion

This preliminary investigation has established the potential for production of SIMFuels by HIP, which would enable the detailed microstructural analysis and dissolution behaviour of more realistic fuels to be studied without the need to handle spent AGR nuclear fuel. The fission product partitioning, phase assemblage, microstructure and porosity of the HIP AGR SIMFuels produced were in good agreement with the available data for AGR spent fuels and SIMFuel produced by conventional reactive sintering, as well as the state of the art knowledge of BWR and LWR spent fuels. It was also established that HIP processing of SIMFuels enables retention of volatile fission products such as Cs and Mo, which should enable more accurate surrogates to be developed in future. Cs was found to be enriched at the SIMFuel / cladding interface, as expected. The observed speciation of CsCl is not inconsistent with some available data on spent fuels, but does not agree with consensus predictions of CsI and Cs_2Te formation. However, in this preliminary study I was not incorporated within the SIMFuel composition. Further development and scale up of this promising approach to SIMFuel manufacture is clearly warranted and is in progress.

5. References

[1] Ewing R C 2015 *Nature Mater.* **14** 252.
[2] International Panel on Fissile Materials, *Managing Spent Fuel from Nuclear Power Reactors – Experience and Lessons from Around the World*, 2011, (eds Feiveson H , Mian Z , Ramana M V , von Hippel F) 1–186.
[3] Strausberg S and Murbach E W 1963 *Ind. Eng. Chem. Process. Des. Dev.* **2** 231.
[4] Lucuta P G, Verall R A, Matzke H and Palmer B J 1991, *J. Nucl. Mater.* **178** 48.
[5] Lucuta P G, Matzke H and Verall R A 1994 *J. Nucl. Mater.* **223** 51.
[6] Sunder S, Shoesmith D W and Miller N H 1997 *J. Nucl. Mater.* **244** 66.
[7] Cobos J, Papaioannou D, Spino J and Coquerelle M 1997 *J. Alloys Compd.* **271** 610.
[8] Wangle T, Tyrpekl M, Cologna M and Somers J 2015 *J. Nucl. Mater.* **466** 150.
[9] Thornber S M, Heath P G, Da Costa G P, Stennett M C and Hyatt N C 2017 *J. Nucl. Mater.* **485** 253.
[10] Hiezl Z, Hambley D I, Padovani C and Lee W E 2015 *J. Nucl. Mater.* **456** 74.
[11] Moore R V, Thorn J D 1963 *J. Brit. Nucl. Energy Soc.* **2** 97.
[12] Morgan S 2011, Size, Grain and Variations, Porosity in Unfailed Spent AGR Fuel, *NNL Memorandum*.
[13] Cohen I and Berman R 1966 *J. Nucl. Mater.* **18** 77.
[14] Rondinella V and Wiss T 2010 *Mater. Today* **13** 24.
[15] Spino J and Papaioannou D 2000 *J. Nucl. Mater.* **281** 146.
[16] Razdan M and Shoesmith D 2013 *J. Electrochem. Soc.* **161** H105.
[17] Desgranges L, Baldinozzi G, Simon P, Guimbretière G and Canizares A 2011 *J. Raman Spectrosc.* **43** 455.
[18] Mohun R, Desgranges L, Jegou C, Boizot B, Cavani O, Canizarès A, Duval F, He C, Desgardin P, Barthe M –F and Simon P , 2019 *Acta Mater.* **164** 512.
[19] Wilbraham R, Rauff-Nisthar N, Boxall C, Howett E, Hambley D, Hiezl Z, Lee W and Padovani C 2018 *Prog. Nucl. Sci. Techno.* **5** 213.
[20] Mignanelli M A and Shaw T L 2009, Information on Phase Chemistry in UK Spent AGR Fuels,

NNL Report, (09) 9834 Issue 1.

[21] Cubiciotti D and Sanecki J E , 1978 *J. Nucl. Mater.* **78** 96.

[22] Pearce J H, Sumerling R and Hargreaves R 1983 *J. Nucl. Mater.* **116** 1.

[23] Walker C T, Bagger C and Mogensen M 1996 *J. Nucl. Mater.* **240** 32.

[24] Lemmens K, Gonzalez-Robles E, Kienzler B, Curti E, Serrano-Purroy D, Sureda R, Martinez-Torrents A, Roth O, Slonszki E, Mennecart T, Gunther-Leopold I and Hozer Z 2017 *J. Nucl. Mat* **484** 307.

[25] Pipon Y, Toulhoat N, Moncoffre N, Jaffrezic H, Gavarini S, Martin P, Raimbault L, Scheidegger A M 2006 *Radiochim. Acta* **94** 705.

Acknowledgement

This research was supported in part by EPSRC under grants EP/N017374/1, EP/S01019X/1, EP/P013600/1 and EP/R006075/1. This work was, in part, performed in the HADES/MIDAS facility at the University of Sheffield, established with financial support of the University of Sheffield, Department of Business, Energy and Industrial Strategy, and the Engineering and Physical Sciences Research Council under grant EP/T011424/1. This research was conducted in part under the auspices of the IAEA co-ordinated Research Programme on Management of Severely Damaged Spent Fuel and Corium (T13015) for which support is gratefully acknowledged.

IOP Conf. Series: Materials Science and Engineering **818** (2020) 012012 doi:10.1088/1757-899X/818/1/012012

Development of phosphate glass and multi-phase titanate ceramic compositions for thermal treatment of irradiated nuclear fuel residues

L Harnett[1], M C Stennett[1], E R Maddrell[2] and N C Hyatt[1*]

[1]Department of Materials Science & Engineering, The University of Sheffield, Mappin Street, Sheffield, S1 3JD.
[2]National Nuclear Laboratory, Sellafield, Seascale, Cumbria, CA20 1PG, UK

*n.c.hyatt@sheffield.ac.uk

Abstract. The highly heterogeneous nature of UK legacy damaged and degraded spent nuclear fuels and so called, 'orphan fuels', prohibits the use of standard conditioning methods. An inventory of UK residual fuels yielded an account for three main fuel types: Magnox, AGR (advanced gas-cooled reactor) and MOx (mixed oxides). A series of glass and ceramic type host systems have been investigated for potential conditioning of these high uranium content spent fuel materials. Electron microscopy and powder X-ray diffraction techniques were used to characterise the prototypical wasteforms. Two sets of low-melt temperature phosphate glass compositions were trialled with additions of CeO_2 to simulate the fluorite structure and large ionic radius of U in oxide fuels. Evolution of monazite-type phases at simulant oxide fuel loadings above 15 wt.% highlighted a potential development into a glass-ceramic hybrid assemblage. Investigation into the use of an alkoxide nitrate synthesis route for SYNROC-F type ceramic precursors has allowed for the demonstration of a sintered host pyrochlore phase containing up to ~40 wt.% fuel simulant CeO_2. Gas evolution has led to increased porosity at higher temperatures and longer sintering times, this may be mitigated by higher pre-calcination temperatures.

1. Introduction

After operation in a nuclear reactor, spent fuel is retained in a storage medium, typically water, to allow for the shielding of workers from the high radiation doses and cooling of decay heat. The UK uniquely differentiates between 'damaged nuclear fuel' and 'failed' fuels: a moderate failure refers to a breach of the fuel element (such as a pinhole), whereas a damaged fuel element has a significant deviation from the typical fuel pin geometry (bent or fractured element). The severity of damaged fuel may confer conditions that can lead to processing and manipulation problems. Failed AGR (Advanced Gas cooled Reactor) fuel has historically been packed into containment using reactor-site facilities, to allow transport to ThORP (Thermal Oxide Reprocessing Plant). Pond storage of some spent fuel has led to degradation and non-conformable cladding geometry; some of this material was still reprocessed with a slightly different procedure, but the stability and integrity of a significant quantity of damaged SNF (Spent Nuclear Fuel) negates use of standard conditioning methods. In 2009, an IAEA working group defined damaged spent fuel as *"any SNF that requires non-standard handling to demonstrate compliance with applicable safety, regulatory or operational requirements"* [1]. Through a review of reactor-fuel specific literature and a collaborative liaison with a National Nuclear Laboratory spent fuel management team, a taxonomy of UK degraded spent fuel and fuel residues of interest was constructed. The major contributors by tons of heavy metal are: Magnox type with some Dounreay Fast Reactor breeder blanket material; low-enrichment type MOx (Mixed Oxide) fuel, steel clad AGR fuel; and a small quantity of experimental carbide fuels.

Many of the spent fuel materials in the UK inventory cannot be processed by the ThORP separation facilities, since activities terminated in 2018, and may be unsuitable for Magnox reprocessing at Sellafield before the end of its operational lifetime in 2020 [2]. Much of this legacy spent fuel is unsuitable for typical vitrification treatments due to volatile content, damage to the fuel pins, contamination through corrosion of containment, novel assembly design, experimental fuel compositions and inclusion of cutting debris [3–6]. The UK is currently considering long term disposal solutions for high activity materials in a mined Geological Disposal Facility (GDF), although very deep boreholes may prove a viable alternative [7].

2. Existing thermal treatment methods

The vast majority of studied and applied vitrification techniques focus on the addition or development of borosilicate glass (either through loading of frit or glass precursors), which can provide high chemical durability and moderate waste loading for ILW (Intermediate Level Waste) streams. However, borosilicate may be unsuitable for some HLW (High Level Waste) and orphan fuels: the poor actinide loading capability may lead to large volume increases of waste, high temperature melts could cause volatilisation of fission products, and the chemical durability may fall short for the particularly long-lived radionuclides [8]. Phosphate glasses were first investigated by the US and Germany during the 1960s, as an alternative to borosilicate: attractive for the low melt temperature and solubility of sulfates and molybdates [9]. Despite these advantages, research slowed as early phosphate glass compositions were unsuitable for waste vitrification due to: highly corrosive melts (therefore propensity to damage melters and refractories); crystallisation during annealing (nucleation of $NaPO_4$ crystallites) [10]; and devitrification at higher temperatures, leading to poor chemical durability [11]. Over the past several decades, research has returned to improving phosphate glasses for actinide waste disposition; additions of Na, Al and Fe oxides are a focus of this work since they may confer better dissolution behaviour when compared to typical borosilicate glass, with complete loading of UO_2 of at least ~12 wt.%, and melting at 1150°C [12]. This research aims to partly build upon recent work that highlighted some chemically durable SAP and SAIP (sodium-aluminium phosphate and sodium-aluminium-iron phosphate) glass compositions that are capable of ~ 49 wt.% UO_3 waste loading. Although annealing did increase crystalline development for the SAIP composition, the leach rate of U was demonstrated to be below 10 g m^{-2} d^{-1}, under PCT-B (Product Consistency Test) conditions [13].

Extensive research and development work has been carried out concerning the use of crystalline ceramic materials for the immobilisation of radioactive waste. A multiphase titanate ceramic, SYNROC (synthetic rock), has been investigated by several nuclear science institutions, in collaboration with international nuclear materials research groups [14]. One compositional variant, known as SYNROC-F was specifically designed for high UO_2 loading to directly immobilise spent fuel. The main components of this formulation are: 95 wt.% cubic U-pyrochlore type similar to betafite ($CaUTi_2O_7$) [15] for disposition of U, rare earths and Sr; up to 5 wt.% hollandite [$Ba(Al,Ti)_2Ti_6O_{16}$] [16] for immobilisation of large ionic radius radionuclides like ^{137}Cs; up to 5 wt.% tetragonal rutile or titanium dioxide (TiO_2) [17]. While synthesis has been carried out using combinations of oxides, precursors generated using sol-gel methods and proprietary techniques, further improvements could be made to streamline the preparation and improve the reactivity of SYNROC precursors, facilitating incorporation of dissolved waste and liquors.

IOP Conf. Series: Materials Science and Engineering **818** (2020) 012012 doi:10.1088/1757-899X/818/1/012012

3. Experimental methodology

3.1. Phosphate glass wasteform

The phosphate glass systems explored were based on two Fe-phosphate compositions: one with small oxide additions of Si, Al and Na, and another only with Na additions. Herein, these compositions will be referred to as IPSAS and SIP, respectively. Oxide and organic constituents were batched and homogenised with the main phosphate base as $NH_4H_2PO_4$ (98% ACS, Alfa Aesar) and Fe_2O_3 (96%, Sigma Aldrich), SiO_2 (99.5%, Alfa), Na_2CO_3 (98%, Alfa), Al_2O_3 (99%, Alfa) and CeO_2 (99.5%, Alfa). CeO_2 additions were used as an inactive simulant for UO_2, as its fluorite structure can provide suitable analogue behaviour prior to approval of active material trials. Batches were homogenised with between 0 – 20 wt.% CeO_2 as simulated spent fuel. The oxide constituents were held for 6 hours at 200°C before 2 hours at a melt temperature of 1250°C under air in an alumina crucible. Glass samples were poured onto a preheated casting plate before annealing for 4 hours at 550°C and slow cooled to room temperature. Samples from the glass coupons were ball-milled for pXRD (powder X-ray diffraction) using a Bruker D2 PHASER diffractometer with a LynxEye detector (Cu Kα, λ = 1.5418 Å radiation). Diffraction patterns were collected in fluorescence mode between $10° < 2\theta < 70°$ with a step size of 0.02° 2θ and dwell time of 3 s per step. Sections of the coupons were also mounted in resin before polishing to a 1 μm mirror finish and coated with a layer of carbon for observation by SEM (scanning electron microscopy). A Hitachi TM3030 electron microscope was coupled with a Bruker Quantax 70 EDX (energy dispersive X-ray) spectroscopy system for chemical analysis.

3.2. Multi-phase titanate wasteform

An alkoxide-nitrate synthesis method was used to generate the reactive titanate precursors. Reagents were sourced from the following: $Ca(NO_3)\cdot 4H_2O$ (99%, Acros Organics); $Ba(NO_3)_2$ (99%, Fluorochem); $Ce(NO_3)_3\cdot 6H_2O$ (99%, Sigma); $Al(OH)_3$ (98%, Acros); and $C_{12}H_{28}O_4Ti$ (96%, Fluorochem). Nitrate components were dissolved in an excess of warm de-ionised water, with the simulant waste added as $Ce(NO_3)_3\cdot 6H_2O$ for co-precipitation and ease of homogenisation. Separately, a mixture of isopropyl alcohol and aluminium hydroxide were agitated before complexing with an equal volume of titanium isopropoxide. The resulting complex was then shear mixed with the dissolved nitrates to ensure rapid and homogenous precipitation. The slurry was dried before calcining for 12 hours at 600°C, the resulting powder was ball milled to increase reactivity before uniaxial compression into pellets and sintering for 1 hour at a range of temperatures 1000°C – 1300°C. Samples were analysed by pXRD using the same instrument as in *Section 3.1* but with adjustment of the detector energy discriminator for rejection of the Fe Kα fluorescence emission from the sample for reduced background, collecting patterns between $10° < 2\theta < 55°$ with a step size of 0.02° 2θ and dwell time of 1 s per step. SEM/EDX was also conducted on the specimens using an identical method as in *Section 3.1.*

4. Results and discussions

4.1. Phosphate glass wasteform

All glasses had a high viscosity during pouring and flowed from upturned crucibles, slowing at a rate proportional to that of the increasing CeO_2 content. A coupon was successfully cast for each composition, and promptly annealed; only the 20 wt.% CeO_2 Na-Fe-phosphate glass fractured in a brittle mode after annealing. Figure 1 shows the X-ray diffraction patterns for IPSAS up to 20 wt.% CeO_2. All of the pXRD patterns demonstrated consistent diffuse scattering for $10° < 2\theta < 40°$ in both systems, indicating a disordered amorphous glass structure. At simulated fuel loadings above 10 wt.% CeO_2, minor developments of crystalline monazite ($CePO_4$, COD 9001646) [18] and cristobalite (SiO_2, COD 1010944) [19] were identified . Figure 2 is an SEM/EDX map of a SIP glass sample batched with 15 wt.% CeO_2. Light grey angular particles with diameters between 5 - 30 μm show elevated Ce detections and inclusion of P surrounding unreacted pure CeO_2. Bulk sample surfaces analysed were largely pristine, with only sporadic inclusion of partially reacted simulant waste and crystallised monazite. The glass matrices of SIP samples had an apparent saturation point for dissolution of Ce, in the case of 15 wt.% waste SIP glass, the bulk composition by EDX had an average saturated Ce content of 9.8 ± 0.1 wt.%. However, the glass matrix local to monazite and CeO_2 crystals revealed a significantly lower average Ce component of 4.8 ± 0.5 wt.%. The bulk glass content for SIP batched with 20 wt.% CeO_2 similarly was found to have a saturated Ce component at 8.9 ± 0.1 wt.%. The waste loadings for IPSAS were not sufficient to confirm a Ce saturation point, although for the IPSAS sample batched with 20 wt.% simulant oxide waste, an average bulk glass composition of 13.1 ± 0.4 wt.% Ce was determined. Further SEM observations on this sample found comparatively less frequent instances of undigested oxide and monazite (as compared to SIP with the same waste loading).

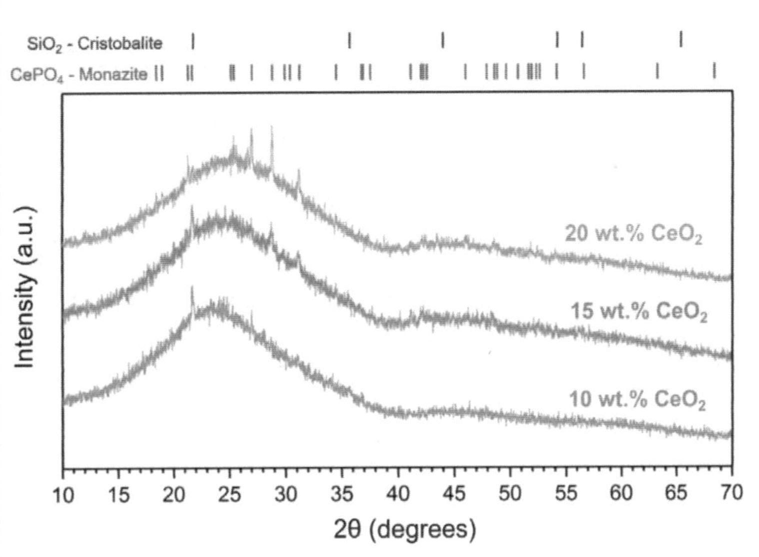

Figure 1. pXRD of IPSAS glasses 10-20 wt.% CeO_2 batches including stick patterns.

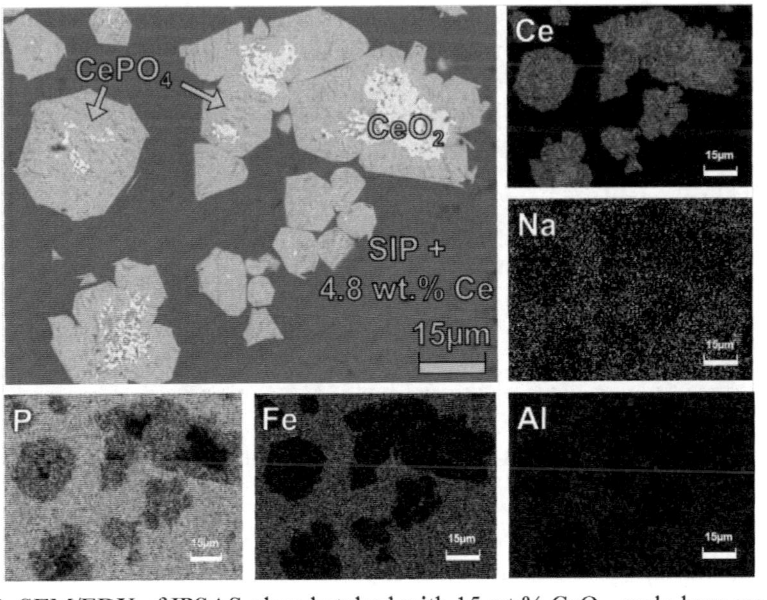

Figure 2. SEM/EDX of IPSAS glass batched with 15 wt.% CeO_2, scale-bars are 15 μm.

IOP Conf. Series: Materials Science and Engineering **818** (2020) 012012 doi:10.1088/1757-899X/818/1/012012

4.2. Multi-phase titanate wasteform

Sintering of the titanate ceramic material afforded significant pellet stability improvements, with increasing apparent hardness and darkening of colour correlating with raising the hold temperature. Figure 3 shows the pXRD patterns for the sintered pellets including data from the 600°C calcined material. The slight diffusivity of the calcined precursor indicates a moderate degree of miscibility and the presence of crystallised CeO_2 (JCPDS 78-0694) is observed. After sintering the material becomes highly crystalline, with a majority pyrochlore phase with peak positions almost identical to the U-containing equivalent pyrochlore $(Ca,Ce)_2Ti_2O_7$ ($[Ca,U]_2Ti_2O_7$, JCPDS 45-1477). A minor balance of crystalline rutile has formed and was detected for material sintered between 1000°C - 1200°C (TiO_2, JCPDS 89-0555); over the same temperature range a concomitant decrease in the relative intensity of TiO_2 and CeO_2 reflections is also evident as the phase assemblage tends towards an apparent equilibrium. However, the diffraction pattern for sample material sintered at 1300°C exhibited what appeared to be some CeO_2 reflections, a minor reflection corresponding to rutile at 40.62° 2θ, and severely diminished reflections for the pyrochlore phase. Figure 4 is a micrograph showing SEM/EDX data for the sample sintered at 1300°C. It was evident here and when analysing the bulk sample, that the light grey contrasted region with a Ce pyrochlore composition is still present as a majority primary phase. This would suggest that the titanate structure has undergone a phase transformation, to that of a new structure-type, with evolution of a gas that has led to the further development of 10 - 20 μm pores. Semi-quantitative analysis using EDX yielded an average pyrochlore composition of $(Ca_{0.57}, Ce_{0.43})_2Ti_2O_7$ (± 0.02 f.u.). Accounting for the oxygen from the CeO_2 incorporated into the structure, the approximate waste loading of this phase is ~40 wt.%. Rutile was also found (dark grey region) as a minor phase with globular morphologies.

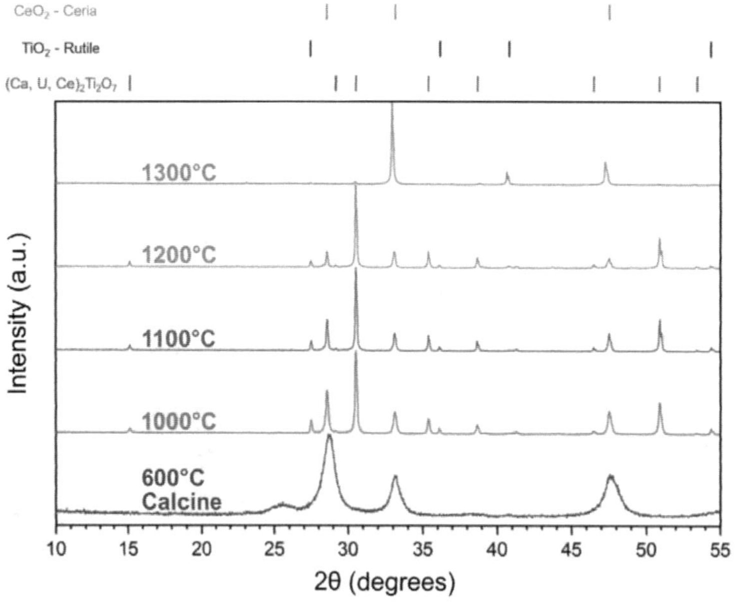

Figure 3. pXRD of co-precipitated multi-phase ceramics, as-calcined and sintered at their respective temperatures, with phase representative stick patterns.

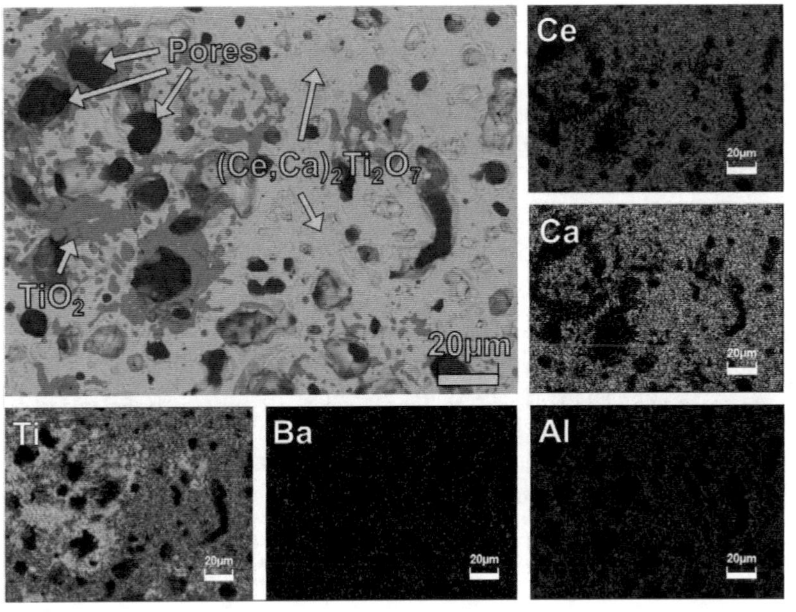

Figure 4. SEM/EDX of SYNROC-F type material cold pressed and sintered at 1300°C for 1 hour. Note Ti, Ba and Ca maps differ from Figure 2, scale-bars are 20 μm.

5. Conclusions and further work

The work conducted was able to successfully demonstrate the incorporation of CeO_2 as a waste simulant up to 13.1 ± 0.4 wt.% in a phosphate glass matrix with a quantity retained in a durable crystalline monazite phase. Further study is necessary to determine the effect of changing melt conditions, such as pressure and temperature, durability assessment by dissolution, and the coordination of Ce and Fe using XAS (X-ray absorption spectroscopy). Up to ~40 wt.% simulant CeO_2 waste was also immobilised into a pyrochlore host phase; further work will be directed at understanding structure evolution above 1200°C, generation of a hollandite secondary phase for minor actinide uptake, and durability of the wasteform.

6. References

[1] IAEA 2009 *Management of damaged spent nuclear fuel No. NF-T-3.6 Guides*
[2] Thompson G R 2014 *Radiological risk at nuclear fuel reprocessing plants*
[3] Jensen S E and Nonbol E 1998 *Description of the Magnox type of gas cooled reactor*
[4] Godfrey H 2017 *Latest understanding of corrosion of Magnox, aluminium and uranium metal wastes in cement*
[5] POST 2000 *Parliam. Off. Sci. Technol.* 1–8
[6] NDA 2011 *Exotic fuels-Dounreay Fast Reactor (DFR) breeder*
[7] NIEA 2009 *Geological disposal facilities on land for solid radioactive wastes guidance on requirements for authorisation*
[8] Sales B C and Boatner L A 1986 *J. Non. Cryst. Solids* **79** 83–116
[9] Terai R, Eguchi K and Yamanaka H 1979 *Cer. in Nucl. Wst. Mgmt.* (USDOE) pp 62–5
[10] Minaev A, Oziraner S and Prokohorova N 1979 *Cer. in Nucl. Wst. Mgmt.* (USDOE) pp 229–32
[11] Jantzen C M 1986 *J. Non. Cryst. Solids* **84** 215.
[12] Mesko M and Day D 1999 *J. Nucl. Mater.* **273** 27.
[13] Stefanovsky S V, Stefanovsky O I, Danilov S S and Kadyko M I 2019 *Ceram. Int.* **45** 9331.
[14] Ringwood A E, Kesson S E, Ware N G, Hibberson W and Major A 1979 *Nature* **278** 219.

[15] Iwata S and Cenzual K 2016 *$Ca_{1.25}U_{0.75}Ti_2O_7$ ($CaUTi_2O_7$) crystal structure.*
[16] Anthony W, Bideaux R, Bladh K and Nichols M 2005 *Handbook of mineralogy - Hollandite.*
[17] Anthony W, Bideaux R, Bladh K and Nichols M 2005 *Handbook of mineralogy - Rutile.*
[18] Yunxiang Ni, Hughes J M and Mariano A N 1995 *Am. Mineral.* **80** 21.
[19] Wyckoff R W G 1925 *Zeitschrift für Krist. - Cryst. Mater.* **62** 1.

Acknowledgement

LH is grateful to EPSRC and the Nuclear Decommissioning Authority for award of a PhD studentship. This research utilised the HADES / MIDAS Facility, established with financial support from the UK Department for Business, Energy and Industrial Strategy and EPSRC under grant reference EP/T011424/1. This research was sponsored in part by EPSRC under grant references EP/P013600/1, and EP/S01019X/1, and conducted in part under the auspices of the IAEA co-ordinated Research Programme on Management of Severely Damaged Spent Fuel and Corium (T13015) for which support is gratefully acknowledged. Statement from the authors: This paper reflects only the author's view and the European Commission is not responsible for any use that may be made of it. No potential conflict of interest was reported by the authors.

IOP Conf. Series: Materials Science and Engineering **818** (2020) 012013 doi:10.1088/1757-899X/818/1/012013

Development of Generic Criteria for Evaluating the Disposability of Thermally Treated Wastes

L Harvey[1], D Galson[1], S Catherin[2], M-A Romero[3], M Fournier[4], A Fuller[1], S Wickham[1]

[1]Galson Sciences Ltd, 5 Grosvenor House, Melton Road, Oakham, LE15 6AX, United Kingdom
[2]Cyclife – EDF Group, 20 Place de la Défense, 92050 Paris La Défense, France (formerly at ANDRA)
[3]ANDRA 1/7 Rue Jean-Monnet, 92298 Châtenay-Malabry, Paris, France
[4]CEA, DEN, DE2D, Univ. Montpellier, Marcoule, France

E-mail: ejh@galson-sciences.co.uk

Abstract. The potential to generate wasteforms with enhanced properties that support safe storage and disposal is one of the key factors driving thermal treatment of radioactive waste. Depending on the specific treatment method and waste substrate, thermal treatment can greatly reduce chemical reactivity, yielding a primary product that is more durable than wasteforms produced via non-thermal routes and with reduced potential for gas generation and other detrimental behaviour in a storage / disposal environment. On the other hand, thermal treatment can concentrate radioactivity into a smaller volume, potentially affecting waste handling and / or classification. It also generates secondary wastes whose management requires consideration as part of a holistic evaluation of thermal treatment. Moreover, some thermal treatment routes do not generate a primary product that is directly disposable without further conditioning.

Generic disposability criteria have been derived that can be used to evaluate the primary products from any form of thermal treatment. These generic disposability criteria highlight the factors that are relevant for waste product disposability and the ways in which thermal treatment can impact on these factors (both positively and negatively). They are equally applicable to any packaging or disposal concept, regardless of the engineered barriers that are present, and in any disposal environment, regardless of its characteristics and the nature of the host rock / geology. They could aid waste management organisations in developing their own disposability criteria, tailored to a particular context, for application in national waste management programmes.

1. Introduction

In order to be disposed of, radioactive wastes must comply with the Waste Acceptance Criteria (WAC) for a disposal facility. WAC identify the characteristics required in a waste product in order to ensure that the waste cannot have a significant detrimental impact on the long-term safety provided by the disposal facility. Compliance with WAC is confirmed by characterisation of the conditioned waste.

Work package 4 (WP4) of the European Commission (EC) THERAMIN project evaluated the disposability of thermally treated waste products and the resulting secondary wastes, considering

IOP Conf. Series: Materials Science and Engineering **818** (2020) 012013 doi:10.1088/1757-899X/818/1/012013

different waste stream / thermal treatment process combinations and the differing disposal concepts and national contexts of countries participating in THERAMIN. It was divided into three tasks [1]:

- *Task 4.1:* Identification and review of criteria and requirements for the disposability of thermally treated waste products. This task considered the relevance of existing WAC, and other requirements relating to the behaviour and performance of waste products, to thermal treatment. Approaches to characterise waste products to determine whether WAC / requirements are met were also defined. The work is reported in THERAMIN Deliverable D4.1 [2].
- *Task 4.2:* Characterisation of thermally treated waste products and secondary waste. Here, characterisation tests have been conducted on thermally treated waste products. Pre-existing data have also been shared. The results are reported in THERAMIN Deliverable D4.2 [3].
- *Task 4.3:* Downstream / Safety Case implications. This task considered the disposability of thermally treated waste based on the outputs from the two previous tasks. Associated deliverables will be published later in 2020.

This paper presents a set of generic disposability criteria that have been derived for thermally treated wastes under THERAMIN Task 4.1 and discusses considerations applicable when measuring compliance with these criteria. It also describes their application in THERAMIN to design a set of characterisation tests for the products from thermal treatment demonstrations, and to feed consideration of disposability into a wider 'value assessment' options evaluation methodology for different thermal treatment routes. The relevance of these disposability criteria for other ongoing EC projects, such as the EURAD ROUTES work package, is also outlined.

2. Generic disposability criteria developed under the THERAMIN project

2.1 Approach to derive generic disposability criteria

The disposability of thermally treated waste products and management of the resulting secondary wastes was examined by reviewing national WAC and other disposability requirements applicable in individual countries (collectively referred to here as 'national disposability criteria') and using these to derive a set of generic disposability criteria that can be applied more generally. Inputs were received from eight countries: Belgium, Finland, France, Germany, Lithuania, Slovakia, Switzerland and the United Kingdom. For each country, a short summary was prepared, setting out factors relevant to the management of radioactive waste and application of thermal treatment in that country, such as the approach to classify radioactive waste, a summary of the inventory, the status of disposal facilities, planned development activities and known issues. Existing criteria applicable to the disposability of thermally treated waste products were then compiled for each participating country. Table 1 summarises the scope of national disposability criteria inputs [2, §4.2].

Table 1. Summary of the scope of national disposability criteria inputs.

Country	Scope of inputs	Type of disposal	Packaging concept	Formal WAC?
Belgium	Planned disposal of short-lived low- and intermediate level waste (LILW) near Dessel in north-east Belgium	Surface	Storage packages grouted in concrete monoliths	No – preliminary – formal WAC will be finalised once a licence for disposal is obtained
Finland	LILW (reactor operating waste) disposal	Near-surface (Olkiluoto VLJ); intermediate-depth (Loviisa VLJ)	Concrete boxes	Yes
France	Long-lived intermediate-level waste (ILW) disposal in clay	Geological (in a clay host rock)	Storage packages grouted in concrete boxes / direct disposal	No – preliminary WAC submitted to the French Safety Authority in 2016

IOP Conf. Series: Materials Science and Engineering **818** (2020) 012013 doi:10.1088/1757-899X/818/1/012013

Country	Scope of inputs	Type of disposal	Packaging concept	Formal WAC?
Germany	Long-lived LILW disposal at Konrad in northern Germany	Geological (host rock to be determined)	Cast iron / steel / concrete boxes	Yes
Lithuania	Planned disposal of long-lived ILW	Geological (host rock to be determined)	Cement-grouted metallic waste	No – preliminary WAC
Slovakia	Low-level waste (LLW) disposal at Mochovce	Surface	Fibre-reinforced concrete packages	Yes
Switzerland	Planned disposal in clay (all radioactive waste)	Geological (clay)	Steel canisters / storage drums in concrete boxes	No – preliminary WAC
United Kingdom	Planned disposal of ILW (also applies to high-level waste/spent fuel disposal)	Geological (salt / clay / crystalline – host rock to be determined)	Waste grouted in steel / cast iron / concrete containers	Not WAC, but formal requirements – specifications to be met for issue of a Letter of Compliance (LoC)

2.2 Observations on national disposability criteria

It is important to note that existing national disposability criteria are not specific to the disposal of thermally treated wastes; they tend to be more generally applicable to all wastes of a certain classification or destined for a particular disposal route. This reflects the relative novelty of thermal treatment as a management route, particularly for LLW and ILW, compared to more conventional waste conditioning approaches such as cement encapsulation, and underpins the need for explicit consideration of criteria for thermally treated wastes. Also, the national inputs provided do not reflect all applicable WAC for radioactive waste disposal in the participating countries. Generally, those provided reflect the wastes that are currently being considered for thermal treatment.

There are some important distinctions between the national inputs received:

- National disposability criteria have been provided for various types of disposal facility, including surface disposal (in the case of responses from Belgium and Slovakia), near-surface disposal (Finland) and geological disposal (France, Germany, Lithuania, Switzerland, United Kingdom). Some criteria will vary depending on the depth of disposal and the safety functions applicable to the wasteform in different disposal contexts.
- For the most part, the national disposability criteria focus on the disposability of long-lived LLW and ILW, although some countries have provided inputs relating to the management of short-lived wastes (Belgium) or inputs that apply to both LLW/ILW and high-level waste (HLW)/spent fuel disposal (Switzerland, United Kingdom). The packaging approach(es) applicable to the different waste types (and reflected in national disposability criteria) depend on the disposal concepts that have been adopted for the relevant disposal facilities in each country, but generally involve disposal in concrete, steel or cast-iron containers.
- Some national disposability criteria have been provided that refer to generic plans for disposal (particularly in the cases of geological disposal in countries that have not yet identified a proposed site or host rock in which to construct a repository). In such cases, the criteria provided are preliminary / provisional, and will be developed further as plans for the disposal facility progress. Other national disposability criteria are site-specific and reflect formal WAC associated with an existing disposal facility (or one that is in the advanced stages of planning).

2.3 Suite of generic disposability criteria

Generic disposability criteria have been derived that can be used to evaluate the products from any form of thermal treatment. These generic disposability criteria highlight factors that are relevant for waste product disposability and the ways in which thermal treatment can impact on these factors (both positively and negatively). They have been developed to be applicable to any packaging or disposal

IOP Conf. Series: Materials Science and Engineering **818** (2020) 012013 doi:10.1088/1757-899X/818/1/012013

concept, for any thermally treated waste, regardless of the engineered barriers that are present, and in any disposal environment, regardless of its characteristics and the nature of the host rock / geology. They are intended to provide a starting point or point of reference for WMOs to tailor their own national disposability criteria to thermally treated waste in a manner that gives confidence that relevant factors are being considered.

The generic disposability criteria have been defined at a relatively high level, focusing on additional requirements and/or clarifications relating specifically to thermal treatment that can build on more widely applicable disposability criteria defined within national programmes. Qualitative, rather than quantitative, metrics are set out, against which disposability can be assessed. This is because numerical requirements identified within national disposability criteria tend to be strongly linked to the national context (e.g. activity limits for different waste classifications), so have limited transferability for wider use outside the country of origin.

Generic disposability criteria and associated considerations are set out in Table 2, based on WAC affected by thermal treatment. Further detail on the underpinning rationale for these criteria is provided in THERAMIN Deliverable D4.1 [2]. If used, these criteria and considerations would need to be applied in conjunction with existing criteria applicable to other wastes that are planned for disposal in a particular facility. They do not stand alone as a complete set of requirements that could underpin WAC for thermally treated waste.

Table 2. Generic disposability criteria specific to thermally treated waste products.

Topic / Category	Generic disposability criterion	Considerations applicable to measure compliance
Dimensions / mass of packages	The dimensions and mass of containers used to package thermally treated waste (and other aspects of the container design) should be compatible with (i) the thermal processing route being employed and (ii) relevant safety functions for storage and disposal, and with all applicable constraints on waste classification, handling, transport and disposal, taking account of the processed waste characteristics.	None.
Package integrity and required lifetime	Apply existing criteria for the disposal context in question. Any additional criteria on package integrity defined for thermally treated waste should be linked to safety functions applied to such waste.	The characteristics of thermally treated waste should be considered as part of demonstrating compliance with existing requirements.
Thermal output	The thermal output of thermally treated waste should not have a detrimental impact on performance of the engineered and natural barriers that make up the disposal system, taking account of the potential for activity to be concentrated during thermal treatment.	None.
Voids	Void space within packages of thermally treated waste should be minimised wherever practicable; this may influence aspects of how thermal treatment is implemented.	None.
Chemical content	Apply existing criteria for the disposal context in question. The choice of thermal treatment route and the design of the associated disposal facility should ensure the chemical compatibility of thermally treated waste with other disposal system components.	None.

IOP Conf. Series: Materials Science and Engineering **818** (2020) 012013 doi:10.1088/1757-899X/818/1/012013

Topic / Category	Generic disposability criterion	Considerations applicable to measure compliance
Chemical durability	Existing requirements on chemical durability for the applicable disposal route should be applied to thermally treated waste. No additional generic disposability criteria for thermally treated waste are considered necessary, although requirements relating to the containment provided by a wasteform may be justified, depending on the post-closure safety case.	If criteria relating to the durability of a thermally treated wasteform are deemed to be required for application in a particular context, then it is recommended that these should be linked to a required containment lifetime (as assumed in the relevant post-closure safety case), rather than to a threshold dissolution rate.
Data management	Data management requirements for the relevant disposal route should be applied to thermally treated waste. In addition, records of the thermal treatment regime applied to the waste should be kept.	None.
Secondary waste	Secondary waste associated with thermal treatment should be minimised to the extent that is practicable.	None.

In a number of other areas against which information on national disposability criteria was gathered, no additional disposability criteria relating specifically to thermally treated wastes are required beyond those already specified for wastes conditioned via non-thermal routes. Moreover, products from thermal treatment may be better able to meet certain existing WAC / requirements as a result of the processes taking place during the thermal treatment step. This is the case for requirements relating to:

- No free liquid/water or gas being present in the waste, since volatile species in the raw waste tend to be driven off.
- No hazardous material being present (and requirements for the waste product to be inert), since reactive waste constituents are typically consumed / destroyed during thermal treatment.
- Minimal gas generation from the waste product, again because of the destruction of reactive waste constituents.
- Production of a robust wasteform, noting that many waste constituents are oxidised / combusted during thermal treatment, ending up in an unreactive form, and the product of thermal treatment is often (although not always) monolithic, with waste species encapsulated or incorporated within a vitreous or ceramic waste matrix.
- Production of a homogeneous wasteform with no localised accumulations of radioactivity.
- Provisions associated with managing the potential for nuclear criticality.
- The wasteform exhibiting mechanical resistance to stresses likely to be imposed during transport, handling and disposal operations.
- Provisions for waste package transport, handling and emplacement, including waste package stacking.
- The waste package activity content.
- The waste package radionuclide inventory, noting that the choice and design of a thermal processing route should be tailored to the radionuclide inventory (and other characteristics) of the raw waste, particularly if thermal treatment is driven by a desire to produce a durable, long-lived wasteform.
- Waste package dose rate limits.
- Surface contamination of the waste package.
- Waste package accident performance, noting that susceptibility to high temperature excursions in the event of a fire is likely to be much reduced, given that the waste has already been subjected to high-temperature treatment.
- Waste package identification / labelling.
- Quality assurance / quality control arrangements.

However, additional considerations relating to the impacts of thermal treatment and the characteristics of the resulting product are often relevant when determining whether existing, more generally applicable criteria have been met. Such considerations include:

- The potential for thermal treatment to concentrate radioactivity and fissile material in the waste product (and thereby to increase heat generation per unit volume).
- The impacts of generating a relatively high density, low voidage wasteform in many cases, which could affect waste package handling.
- The potential for thermal treatment to introduce new mechanisms for contamination of equipment and/or waste packages (e.g. splashing, particulate generation and/or carry-over to the off-gas system).
- The chemistry and mechanical properties of thermally treated wastes, which may behave differently during handling, storage and disposal, thereby introducing new uncertainties relating, for example, to chemical compatibility in a disposal environment.
- Safety functions applicable to thermally treated wastes may or may not be the same as those applicable to other wastes to be disposed of in the same facility, depending on the drivers for implementing thermal treatment. Evaluation of thermally treated waste in the post-closure safety case for a disposal facility may therefore differ from that for other wasteforms.

3. Design of characterisation tests for products of thermal treatment

Having collated national disposability criteria and derived generic disposability criteria, an exercise was conducted to identify WAC requiring characterisation of thermally treated waste products in order to test compliance, along with identification of suitable analytical techniques to meet these characterisation requirements. Table 3 summarises criteria identified as requiring characterisation tests to be performed, and identifies physico-chemical properties that can be measured to evaluate whether each of these criteria have been met, along with some examples of applicable, commonly used, measurement techniques. It can be seen that most criteria can be verified through the use of electron microscopy, possibly combined with analyses of the chemical and radiological composition of the sample. However, some WAC require more specialist techniques to measure compliance.

Table 3. Measurable properties and techniques applicable for verifying whether WAC have been met.

Waste Acceptance Criterion	Properties	Examples of Applicable Techniques
No, or limited, free liquid or gas and limited potential for gas generation	Homogeneity of the waste	Thermogravimetric analysis (TGA), X-ray fluorescence (XRF), electron microscopy
Permeability and/or diffusivity of the waste sufficient to enable gas to escape without affecting wasteform integrity	Permeability + diffusivity	XRF, electron microscopy
No or limited content of hazardous materials (combustible, pyrophoric, reactive, etc.)	Homogeneity of the waste (no untreated area) + identification of chemical species in the waste	XRF, X-ray diffraction (XRD), Inductively coupled plasma (ICP) analysis after dissolution
Immobilisation of radionuclides, including homogeneity and passive safety under both normal and 'accident' conditions.	Distribution of radionuclides in the waste	α-spectrometry, autoradiography, Raman spectroscopy
Limited voids / limited porosity (e.g. to limit the potential for collapse of waste package stacks in a disposal facility)	Porosity	Wide-angle X-ray scattering (WAXS), surface area measurement by gas physisorption, porosimetry
No localised accumulations of radioactive material	Homogeneity of the waste / microstructure	XRF, electron microscopy

Waste Acceptance Criterion	Properties	Examples of Applicable Techniques
Chemical durability of the waste product	Chemical durability (leaching behaviour)	Leaching tests, ICP, ion chromatography, UV-Visible spectroscopy, α-spectrometry
Mechanical resistance of the waste product to stresses imposed during transport, handling and disposal operations	Mechanical behaviour	Mechanical resistance test methods (compression, tension,)
Limited or no metal, e.g. in Belgium: no metal with a redox lower than 0.84 V Standard Hydrogen Electrode (SHE) (specified because of the potential for such metals to corrode and produce significant quantities of hydrogen gas).	Homogeneity of the waste / microstructure	XRF, electron microscopy, optical microscopy
Thermal conductivity of the waste product (especially for self-heating waste)	Thermal conductivity / thermal behaviour	Thermal conductivity measurement

This exercise was used as the basis for planning characterisation tests conducted under THERAMIN Task 4.2. The results of these tests are reported in THERAMIN Deliverable D4.2 [3] and are the subject of several other papers published elsewhere in this issue of *IOP Conf. Ser.: Mater. Sci. Eng.*

Such tests would only need to be conducted on selected products from an industrial-scale thermal treatment facility, assuming that quality controls guarantee the representativeness and transferability of tests conducted on sample materials.

4. Disposability inputs to THERAMIN value assessment

A value assessment approach has been developed under THERAMIN WP2. It provides a structured methodology for evaluating the potential application of a thermal treatment technology to treat a waste stream of interest. 'Value' in this context is defined as realisable benefit in safety, monetary and/or environmental outcomes from implementing an option at a specified time. This includes benefits and challenges across all stages of the waste management lifecycle. The value assessment methodology aims to integrate learning from across the THERAMIN project.

The methodology developed employs a multi-attribute assessment approach that builds on the UK Nuclear Decommissioning Authority's (NDA's) value framework [4]. The following assessment attributes are included in the THERAMIN value assessment methodology:

- Operational and transport safety
- Environment impact
- Impact on disposability and long-term safety
- Factors affecting the implementation of a thermal treatment technology
- Timescales
- Technical readiness of the technology
- Strategic cost impacts

Each of these attributes is sub-divided into a series of 'data categories', accompanied by a set of assessment considerations that provide guidance to the user on factors to consider when conducting a value assessment. Two data categories are defined under the impact of disposability and long-term safety attribute: the ability to meet applicable WAC and the disposability of secondary waste. The generic disposability criteria described in this paper are the principal suite of assessment considerations relating to these data categories, alongside identification of potential disposal routes for secondary waste. Further information on the THERAMIN value assessment methodology is provided in [5].

5. Links to other EC projects

Several other collaborative EC projects are considering, or have recently considered, the development and application of WAC within their work programmes, albeit with differing areas of interest and emphasis (and not focused on thermal treatment as in THERAMIN). There are opportunities for synergy across projects and a need to ensure future project activities learn from recent developments.

The EC Horizon 2020 project **"CHANCE"** (Characterization of Nuclear Waste for its Safe Disposal in Europe) aims to address issues associated with the characterisation of conditioned wastes. It aims to establish, at the European level, a comprehensive understanding of current conditioned radioactive waste characterisation and quality control schemes across national radioactive waste management programmes, and to further develop, test and validate techniques that will improve the characterisation of conditioned radioactive waste, particularly those that cannot easily be studied using conventional methods. Knowledge of applicable WAC is an important input to CHANCE in order to identify characterisation requirements, and was a focus area for a recent information gathering exercise [6].

WAC are also considered within WP9 **("ROUTES")** of the European Joint Research Programme on Radioactive Waste Management (EURAD), which commenced in June 2019 and runs for five years. ROUTES focuses on waste management routes in Europe from cradle to grave. It aims to:

- Provide an opportunity to share experience and knowledge.
- Identify safety-relevant issues and their research and development needs.
- Compare characterisation, treatment, conditioning and long-term waste management strategies, and identify opportunities for collaboration between European Union (EU) Member-States.

Task 4 of ROUTES involves identification of WAC used in Member-States for different disposal alternatives in order to inform development of WAC in countries without WAC / disposal facilities. Countries will also share experience of managing challenging wastes without a final disposal solution.

Both WAC and thermal treatment will be considered under the auspices of a new (still to be approved) EC project **"PREDIS"** (Pre-disposal Management of Radioactive Waste), which is expected to run for four years from June 2020. The objectives of PREDIS include developing new solutions for treatment and conditioning of waste for which no industrially mature or adequate solution is currently available, and improving existing solutions with safer, cheaper or more effective processes. Work packages will focus on metallic, liquid organic, solid organic and cemented wastes, as well as considering wider strategic implementation of treatment and conditioning solutions. The wider investigation of generic WAC also forms part of PREDIS.

Acknowledgements
This project has received funding from the Euratom research and training programme 2014-2018 under grant agreement No 755480. This paper reflects the authors' views and the European Commission is not responsible for any use that may be made of it.

References
[1] Scourfield S J, Kent J E, Wickham S M, Nieminen M, Clarke S, Frasca B 2020 Thermal treatment for radioactive waste minimisation and hazard reduction: overview and summary of the EC THERAMIN project, *IOP Conf. Ser.: Mater. Sci. Eng* same issue

[2] Andra *et al.* 2018 *Waste Acceptance Criteria and requirements in terms of characterisation*, European Commission THERAMIN Deliverable D4.1.

[3] Andra *et al.* 2019 *Characterization of thermally treated waste products*, European Commission THERAMIN Deliverable D4.2.

[4] Nuclear Decommissioning Authority 2016 *The NDA Value Framework*, Version 1.2.

[5] Fuller A, Doudou S, Kent J, Harvey L, Wickham S 2020 *Assessing the value of thermal treatment technologies, IOP Conf. Ser.: Mater. Sci. Eng* same issue.

[6] Bucur C, Dodaro A, Ferruci B, Meskens G, Ricard D, Tietze-Jaensch H, Thomas P and Turcanu C 2019 *Synthesis of commonly used methodology for conditioned radioactive waste characterization*, European Commission CHANCE Deliverable D2.2.

IOP Conf. Series: Materials Science and Engineering **818** (2020) 012014 doi:10.1088/1757-899X/818/1/012014

Characterisation of co-mixed HIP wasteforms for Magnox sludge and clinoptilolite wastes

S A Walling[1], L J Gardner[1] and N C Hyatt[1]*

[1]Department of Materials Science & Engineering, The University of Sheffield, Mappin Street, Sheffield, S1 3JD.

*n.c.hyatt@sheffield.ac.uk

Abstract. Co-mixed simulant wasteforms consisting of calcined Magnox sludge simulant and clinoptilolite, with additions of a glass forming frit and CeO_2 or U_3O_8 were processed using Hot Isostatic Pressing (HIP). This enabled the production of high waste loaded materials, with the successful incorporation of both simulant and active material. These formed heterogeneous glass-ceramic products, with decomposition of raw materials and some vitreous phase formation. The aqueous durability of these materials was assessed over a 28-day period using a modified PCT test, and favourably compared to the durability of an international glass. Overall this verifies the potential for HIP technology to be used in wasteform production, with potential large reductions in waste volume, especially if co-mixed wastes are considered.

1. Introduction

The safe conditioning and disposal of nuclear wastes within durable materials is of paramount importance to safeguard current and future generations. Past and present nuclear power generation within the UK has resulted in a legacy of different waste streams, many with unique or challenging chemistries and physical characteristics. Two of the waste streams of interest within the EC THERAMIN project (Thermal Treatment for Radioactive Waste Minimization and Hazard Reduction) are Magnox sludge and spent clinoptilolite (an ion-exchange material). The former of these is a magnesium rich sludge, predominantly resulting from the corrosion of Magnox fuel cladding within various ponds and silos across the Sellafield site. This is predominantly $Mg(OH)_2$, with quantities of corroded uranium [1-2]. The clinoptilolite is a naturally occurring heulandite aluminosilicate zeolite used to selectively remove Cs and Sr from waste waters at the SIXEP plant prior to discharge [3-4]. Both of these waste streams are of interest for thermal treatment for different reasons. Magnox sludge is a high volume material, however de-watering of sludge and further calcination would result in a much lower volume of waste if a suitable thermal treatment could be applied. Clinoptilolite on the other hand is chemically very suitable to use as a starting point for glass formation, being rich in Si and Al, in addition to Na, Ca and K. This opens up the potential for co-processed wastes, with a lower amount of glass-forming additives required during thermal treatment, and consequential higher waste loading.

The objective of this work was to assess the suitability of co-processed Magnox sludge simulant and clinoptilolite, using Hot Isostatic Pressing (HIP) technology. This is a batch process within sealed steel canisters, which are subjected to both heat and pressure to form consolidated waste packages. This in-canister thermal treatment greatly limits the loss of volatile species during processing, while able to produce a glass, glass-ceramic or fully ceramic material. Here we have produced both an inactive (using CeO_2) and an active (using U_3O_8) HIP wasteform, both of which have been characterised and subject to aqueous dissolution to determine their durability and therefore, the suitability of managing co-mixed wastes by HIPing.

2. Materials and methods

Two small HIP canisters were prepared, each containing ~50g material (Table 1). These were to simulate co-mixed wastes, with a calcined Magnox sludge simulant (CMgS – a corroded magnesium simulant prepared by NNL), clinoptilolite (Clino), and a glass frit (MW ½ Li glass, a simple sodium lithium borosilicate glass used as the base constituent for UK HLW vitrification) which was added to assist glass formation within the wasteform. To one of these, CeO_2 was added as a uranium surrogate, to the other U_3O_8 was added on an equimolar basis. The CMgS was calcined at 950 °C for 3 hours to remove water (and will result in conversion of $Mg(OH)_2$ to MgO), substantially reducing the volume in the process. Both samples achieved a waste loading of ~90%, based on dry materials. Clinoptilolite was similarly treated to remove any excess water (also likely to induce amorphisation). The MW ½ Li glass is supplied as a broken frit with the composition (wt. %) SiO_2 63.4%, B_2O_3 22.5%, Na_2O 11.4%, Li_2O 2.7%.

Table 1. Composition of radioactive HIP wasteforms

Can No.	CMgS (g)	MW ½ Li (g)	Clino (g)	CeO_2 (g)	U_3O_8 (g)	Waste loading (wt. %)*
HIP – CeO_2	16.57	5.03	28.12	0.30	-	89.94
HIP – U_3O_8	16.51	5.02	28.01	-	0.48	90.84

*where waste loading refers to CMgS, clinoptilolite and CeO_2/U_3O_8 content

To reduce any further pre-processing, the materials were loosely mixed together, and no milling or size reduction was undertaken. This will likely result in a heterogeneous material, however additional processing of real wastes would increase complexity and produce secondary effluents, which then require disposal. For each canister, the dry precursors were batched, and packed into a HIP canister (15 cm^3 volume, 304 stainless steel) using a hydraulic press to enhance packing density (loosely packed material may result in severe can deformation (i.e. uncontrolled final shape) during the HIP process due to material consolidation). The lid was welded, can placed under vacuum while heated to 600 °C to remove any remaining water, then the evacuation tube was crimped and welded to fully seal the canister.

Both canisters were processed using an American Isostatic Press AIP-630H using an Active Furnace Isolation Chamber (AFIC) to ensure safe containment of radioactive material under any mis-operation conditions. HIP canisters were heated to 1250 °C over 260 minutes, then held at temperature for 120 minutes, before cooling to room temperature at 10 °C min^{-1}. Varying ramp rates and several hold points on the ramp up were programmed to control canister deformation, overall averaging ~4.7 °C min^{-1}. Pressure was increased (using argon) commensurate with temperature, with maximum pressure held during the dwell at maximum temperature. The CeO_2 can was held at an average of 87 MPa, with the U_3O_8 can held at an average of 70 MPa. Both these were lower than the target pressure due to a small leak in the compressor and a vent valve, but this is unlikely to affect the final phase assemblage,

Post HIP processing the samples were sectioned with a diamond wafering blade using a Buehler Isomet 1000 precision saw, half the can was mounted into resin, then ground and polished to a 1 μm finish for scanning election microscopy / energy-dispersive X-ray spectroscopy (SEM/EDX), with the other half crushed for X-ray diffraction (XRD) analysis. Powder X-ray diffraction patterns were collected between $5° < 2\theta < 70°$ using a Bruker D2 Phaser diffractometer with a Lynxeye detector and a Ni filtered Cu K_α radiation (1.5418 Å) using a step size of 0.02° 2θ and a count time of 3 s per step. Micrographs were collected using a low vacuum Hitachi TM3030 analytical benchtop SEM with an integrated Bruker EDX system (Quantax 70) at 15 kV and a 7.0 mm working distance.

Samples were subjected to short aqueous durability testing using a modified product consistency test (PCT) leaching methodology based on ASTM C1285 (but using a larger particle size). This utilised 10 mL of Type 1 (deionized) water in PFA pots, with sample material crushed and sieved to between 125-250 µm. A surface area / volume of 10 m^{-1} was used, based on geometric surface area, with static leaching at 90 °C and sampling at regular intervals up to 28 days. Aliquots of leachant were taken and filtered using 0.2 µm cellulose acetate filters, pH measured, then acidified using ultrapure HNO_3 and analysed by inductively couple plasma optical emission spectrometry (ICP-OES, Thermo Fisher 6000 iCAP Duo). Normalised mass loss for elements was calculated using data obtained via X-ray fluorescence (XRF) analysis (Table 2), and compared against a well-known international simple glass (ISG)

Table 2. XRF composition of HIP – CeO_2 and HIP – U_3O_8 post processing and, ISG (wt. %)

	HIP – CeO_2	HIP – U_3O_8	ISG [5]
Al_2O_3	7.52	7.96	5.9
B_2O_3	2.23	2.11	17.7
CaO	1.06	1.04	4.6
CeO	0.44	-	-
Fe_2O_3	0.76	1.02	-
K_2O	2.88	3.10	-
Li_2O	0.28	0.30	-
MgO	25.64	23.30	-
Na_2O	1.98	2.16	12.4
SiO_2	55.68	56.30	56.3
TiO_2	<0.10	0.16	-
U_3O_8	-	1.12	-
ZrO_2	<0.10	0.10	3.2

3. Results and discussion

3.1. Wasteform characterisation

Both HIP canisters were successfully processed, with material consolidation and corresponding slight reduction in volume. Photographs of both canisters are shown in Figure 1, both before and after HIP processing. The processed material is quite heterogeneous, but have both formed solidified products with little deleterious interaction with the HIP canister.

X-ray diffraction analysis of the crushed materials reveals various crystalline phases. For the CeO_2 sample in Figure 2, forsterite ($MgSiO_3$), ceria (CeO_2), periclase (MgO) and quartz (SiO_2) were readily identifiable, additionally a feldspar mineral was tentatively assigned, potentially albite ($NaAlSi_3O_8$) or anorthoclase ($(Na,K)AlSi_3O_8$). Several reflections remain un-assigned, though some of these do correspond to kotoite ($Mg_3(BO_3)_2$), this could not be assigned in Figure 2 with confidence without further evidence. The diffraction pattern for the U_3O_8 sample (Figure 3) was very similar, identifying the same crystalline phases (except, obviously no CeO_2). This sample exhibited a high background signal (especially the diffuse scattering region between ~15-25° 2θ) from Kapton tape, which was required to minimise loose active powders. Again several reflections remain unassigned, with kotoite ($Mg_3(BO_3)_2$) a possible fit for several of the reflections but without further evidence we are unable to commit to this assignment. An additional reflection was evident at 15° 2θ, which could not be assigned. This might fit anhydrous sodium diuranate ($Na_2U_2O_7$), though little evidence was found for this in further analysis (SEM/dissolution).

IOP Conf. Series: Materials Science and Engineering **818** (2020) 012014 doi:10.1088/1757-899X/818/1/012014

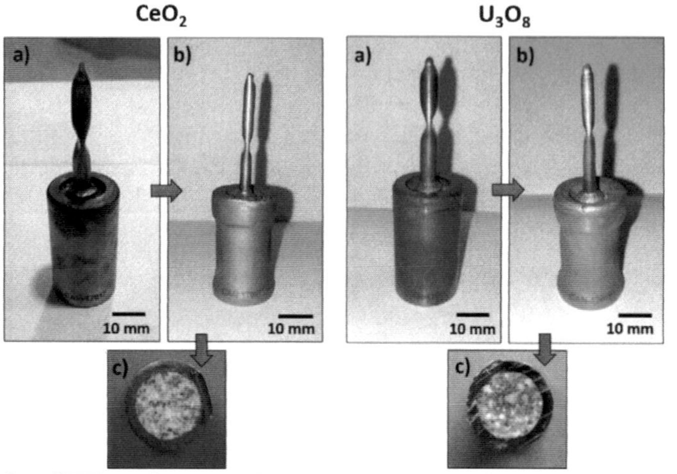

Figure 1. Photographs of HIP canisters a) after packing and welding, b) post-HIP, c) sectioned canister cross-section.

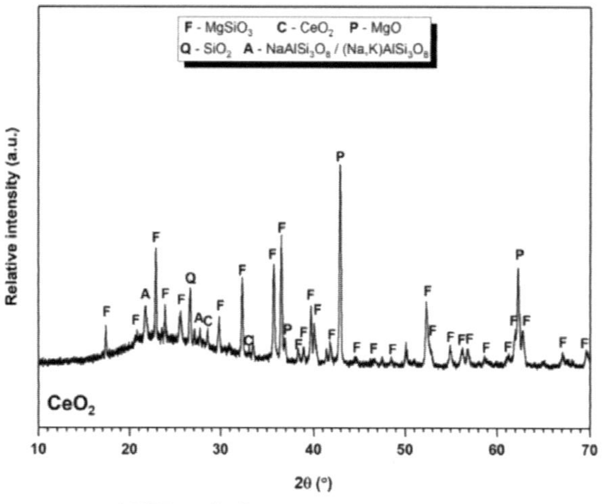

Figure 2. X-ray diffraction pattern of HIP – CeO_2.

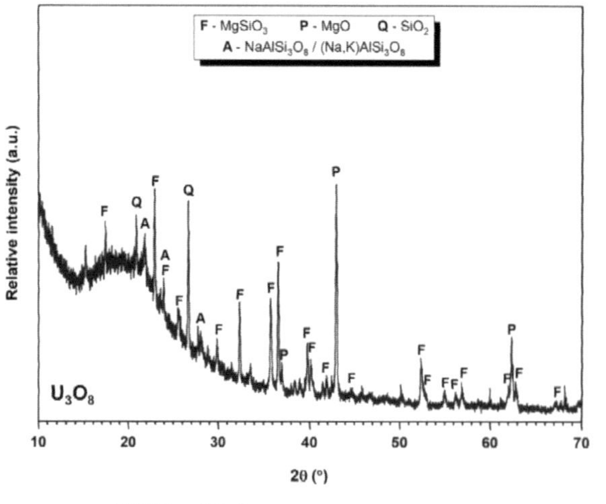

Figure 3. X-ray diffraction pattern of HIP – U_3O_8.

Scanning electron microscopy of the product wasteforms reaffirmed the heterogeneous nature of these materials (Figure 4). Two levels of magnification are presented, alongside elemental mapping to aid phase identification. Both samples were broadly similar in microstructure, with the difference in cracking likely due to damage obtained while cutting the HIP canister or grinding/polishing.

In HIP – CeO_2, region 'a' is likely a partially vitrified clinoptilolite grain (given no clinoptilolite was identified via XRD, it has likely decomposed). This was evidenced by the presence of Ca, Al, Si and Na together (clinoptilolite is nominally $(Na_{0.5}K_{2.5})(Ca_{1.0}Mg_{0.5})(Al_6Si_{30})O_{72}{\cdot}24H_2O$)). Region 'b' is likely an Mg-Na-Si based glass, inside which unreacted MgO particles were visible, labelled as region 'c'. Presence of Na and Ca within the MgO is likely due to background signal. The bright white region 'd' corresponds to unreacted CeO_2.

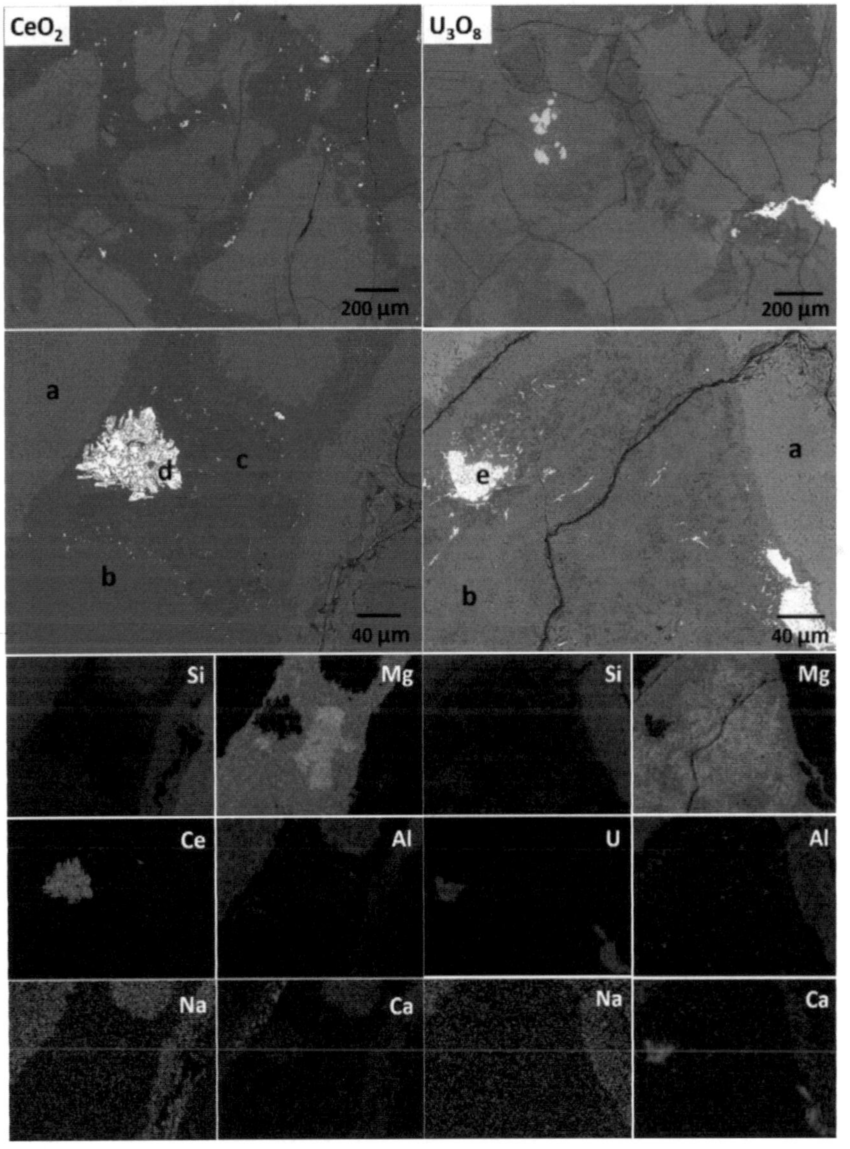

Figure 4. SEM/EDX of HIP – CeO_2 (left) and HIP – U_3O_8 (right).

For HIP – U_3O_8, the sample exhibited much the same microstructure as the CeO_2 sample with Ca, Al, Si and Na rich region 'a', and a Mg-Na-Si glassy phase in region 'b'. Unreacted MgO is likely present in the centre of the micrograph, evidenced by the more intense Mg EDX signal, however the boundary between this and the Mg-Na-Si glass is not distinct. Region 'e' is of interest, as this initially appeared to be unreacted U_3O_8, however upon further analysis, there was also a strong signal from Ca within this region. This could be the presence of a calcium uranate phase, potentially also accounting for one or more of the unassigned XRD reflections (Figure 3).

3.2. Chemical durability assessment

Although these materials are heterogeneous, it is the overall bulk dissolution of these materials against which we benchmark their durability. The results of the 28-day static dissolution tests are shown in Figure 5 with the normalised mass losses for respective elements from the two wasteforms. These have been compared against the International Simple Glass (ISG), which was tested under the same experiment conditions as an in-house benchmark of durability. Although it should be noted that ISG is a simulant high-level waste glass, not an intermediate- or low-level glass, and as such the required characteristics are different for simulant wasteforms than the wasteforms produced in this study.

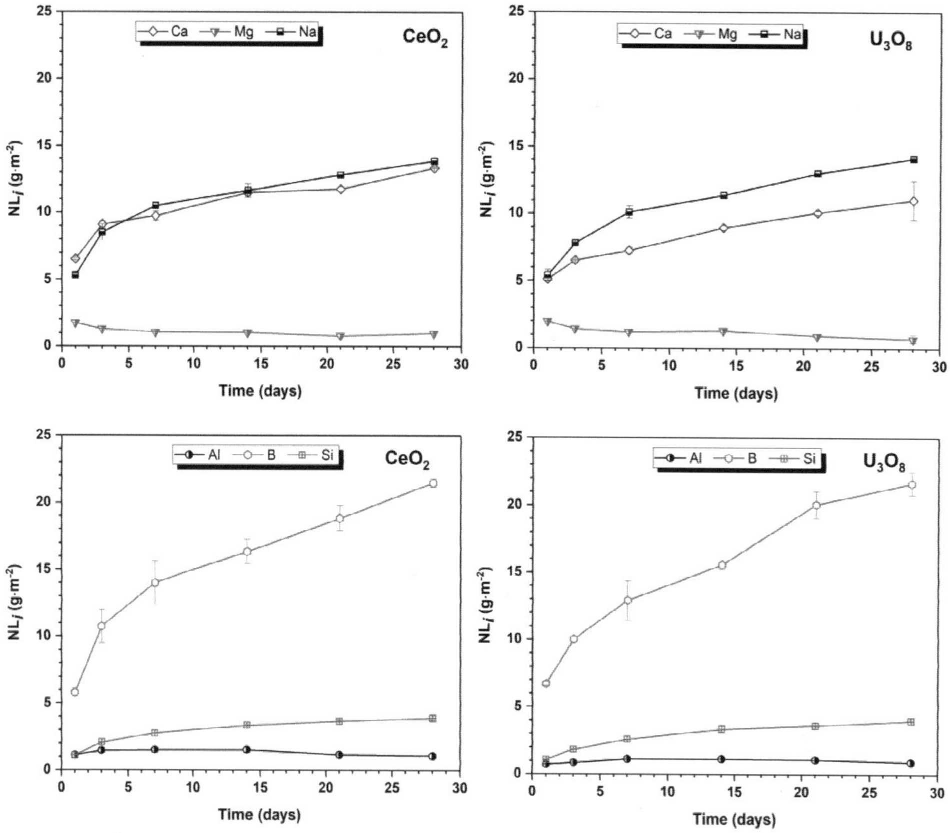

Figure 5. Normalised elemental mass loss (NL_i) for HIP – CeO_2 and HIP – U_3O_8 up to day 28.

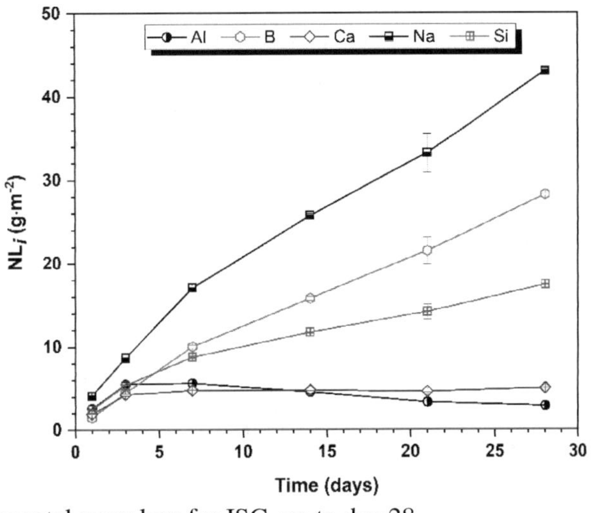

Figure 6. Normalised elemental mass loss for ISG, up to day 28.

During dissolution the recorded pH values for the two samples followed the same trend. For HIP – CeO_2 the pH at 1 day was 9.8, decreasing to 9.2 after 28 days. For HIP – U_3O_8 the pH at day 1 was 9.8, decreasing to 8.9 after 28 days. Alongside this, there is immediate elemental release within the first day for all elements except Ce. Initially Mg and Si were the highest concentration (in mg/L, however the composition is >80 wt.% MgO + SiO_2), normalising to elemental fraction B, Na and Ca were the highest released elements at 5-7 g·m^{-2} for both CeO_2 and U_3O_8 HIP samples. For the CeO_2 sample; B, Na and Ca continued to be released up to 28 days leaching with final release rates of 21.5 ± 0.3 g·m^{-2} for B, and 13.9 ± 0.2 g·m^{-2} for Na at 28 days. Both Al and Si release were slow, with Al slowly decreasing after the first few days, and Si only slowly increasing. The final Si normalised mass loss (3.9 ± 0.3 g·m^{-2} compared to 17.0 ± 0.1 g·m^{-2} in ISG), might indicate comparable higher durability, along with a slightly reduced B normalised mass loss (21.5 ± 0.3 g·m^{-2} compared to 28.2 ± 0.1 g·m^{-2} in ISG)

The U_3O_8 sample follows a very similar trend, with only minor deviations (notably, a lower Ca release rate). The final release rates for B and Na are 21.6 ± 0.9 g·m^{-2} and 14.1 ± 0.1 g·m^{-2} respectively, with final Si normalised mass loss of 4.0 ± 0.1 g·m^{-2}. These are all very close to the release rates determined for the CeO_2 sample, as expected given the close batch composition and phase assemblages. Though we have compared these samples here to the ISG a benchmark, it should be noted that ISG is not in itself a proposed wasteform, instead representing a simplified six-oxide simulant HLW glass for assessing dissolution mechanisms. Their structures are very different, as are their chemistries, however ISG does represent a vitrified product within the envelope of actual vitrified HLW. A well-recognised glass was required for comparison to determine whether, using this leaching methodology, these wasteforms perform adequately.

The relatively similar leaching characteristics of the U_3O_8 sample compared to the CeO_2 sample corroborate the XRD and SEM analysis that the bulk phase assemblage and microstructure of these materials were similar, and also that no readily soluble Ce or U bearing phases were formed (Ce release was below the ICP-OES limits of detection, while the U normalised mass loss was 1.2 ± 0.2 g·m^{-2} at 28 days).

4. Conclusions

Two co-mixed waste simulants containing calcined Magnox sludge simulant and clinoptilolite have been successfully processed using HIP technology, both with CeO_2 as a uranium surrogate, and using U_3O_8. Similar phase assemblages and dissolution characteristics were reported for both wasteforms, though with differences in the final Ce/U assemblages. The clinoptilolite crystal structure was destroyed, forming vitreous phases in-between a Mg-Na-Si glassy matrix containing the CeO_2 / calcium-uranium phases. The chemical durability of these materials was favourable over a 28-day period, though with varying elemental release rates given the heterogeneous nature of the materials produced. Ultimately a high waste loading has been achieved with these materials, demonstrating the possibility of co-mixed wastes, and was able to successfully incorporate uranium – though this process does require the pre-calcination of waste streams to ensure suitability for the HIP process. More widely, the results presented here demonstrate the potential efficacy of hot isostatic pressing for the co-treatment of sludge and spent inorganic ion exchange materials present on the Fukushima Dai-ici and other legacy decommissioning sites.

5. References

[1] Nuclear Decommissioning Authority 2019 *Waste Stream 2D95.1: Magnox Fuel Storage Pond Sludge* (Moor Row, Cumbria: NDA).
[2] Gregson C R, Goddard D T, Sarsfield M J and Taylor R J 2011 *J. Nucl. Mater.* **412** 145.
[3] Nuclear Decommissioning Authority 2019 *Waste Stream 2D26: Ion Exchange Material (Clinoptilolite) and Sand* (Moor Row, Cumbria: NDA).
[4] Howden M 1987 *Proc. Inst. Mech. Eng., Part A* **201** 1-15.
[5] Kaspar T C Ryan J V, Pantano C G, Rice J, Trivelpiece C, Hyatt N C, Corkhill C L, Mann C, Hand R J, Kirkham M A, Crawford C L, Jantzen C M, Du J, Lu X, Harrison M T, Cushman C, Linford M R and Smith N J, 2019 *npj Mater. Degrad.* **3** 15.

Acknowledgement

The research leading to these results has received funding from the European Union's Horizon 2020 Euratom research and innovation programme under Grant Agreement No 755480 – Theramin and UKRI / EPSRC under grant reference EP/N017617/1. The HIP active upgrade was funded in-part by EPSRC under grant EP/L014041/1 Decommissioning, immobilisation and storage solutions for nuclear waste inventories (DISTINCTIVE). This research utilised the HADES / MIDAS facility at The University of Sheffield established with financial support from EPSRC and BEIS, under grant EP/T011424/1. The authors would like to acknowledge the National Nuclear Laboratory for supply of inactive precursor materials and for continued collaboration.

IOP Conf. Series: Materials Science and Engineering **818** (2020) 012015 doi:10.1088/1757-899X/818/1/012015

Incineration-vitrification of a mixture of zeolites, diatoms and ion exchange resins using the SHIVA process

M Fournier and F Rousset

CEA, DEN, DE2D, Univ. Montpellier, Marcoule, France

maxime.fournier@cea.fr

Abstract. The Advanced Incineration-Vitrification Hybrid System (SHIVA) process is well suited to treat organic and mineral waste, with high alpha contamination management capabilities. It allows, in a single reactor, waste incineration by a plasma burner and ash vitrification in a cold wall direct glass induction melting system. The SHIVA trial demonstrates the successful processing of an inactive waste stream analogue that contains a mixture of mineral and organic ion exchange media. The waste loading attains a promising value of 38 wt.% and a glass wasteform is obtained. Further, the microstructure, chemical composition, and chemical durability of the wasteform are characterised.

1. Introduction

The Thermal treatment for radioactive waste minimisation and hazard reduction (THERAMIN) project is a European Commission programme, which is jointly funded by the Horizon 2020 Euratom research and innovation programme and various European nuclear waste management organizations (WMOs). The THERAMIN project is being conducted between June 2017 and May 2020. Twelve European WMOs and research and consultancy institutions from seven European countries are participating in THERAMIN. The overall objective of THERAMIN is to provide improved, safe, long-term storage, and disposal of intermediate as well as low-level wastes that are suitable for thermal processing. The work programme provides a vehicle for coordinating research in the entire European Union and demonstrating technologies, which are designed to improve the understanding and optimisation of the application of thermal treatment in radioactive waste management programmes across Europe. Additionally, this project will facilitate the development of technologies to increase their Technology Readiness Level (TRL).

In the framework of the THERAMIN project, studies were carried out by the French Alternative Energies and Atomic Energy Commission (CEA) on the incineration-vitrification of mixed inactive waste, which contains inorganic and organic ion exchange media using the Advanced Incineration-Vitrification Hybrid System (SHIVA) process. SHIVA combines heating methods by employing transferred arc plasma and direct induction; the waste incineration is carried out above the plasma torches and the molten glass is heated by the induction generator.

2. Description of the SHIVA process

The SHIVA process (Figure 1) uses cold wall direct glass induction melting and plasma burner technologies. It is therefore an incineration-vitrification process well suited for organic wastes.

SHIVA consists of a water-cooled, stainless steel cylindrical reactor, which is equipped with a flat coil at the bottom and a transferred arc plasma system in the reactor chamber. The SHIVA bottom structure is built to be transparent to the electromagnetic field such that the glass is directly heated by the field, while the cylindrical shell is not. The plasma torches were developed to minimise their maintenance requirements, in particular, the graphitic consumable electrodes are fed automatically. As oxygen is used and no secondary chamber is present, the gas treatment is simple and compact. It consists of an electrostatic tubular filter and a gas scrubber. The dust in the filter is recovered in a bottom ashtray for recycling. The glass is drained from the cold crucible.

SHIVA was designed to treat organic waste, potentially with a high chloride or sulphur content. The waste can be in solid or liquid form. SHIVA is not compatible with metallic waste nor does it have a separate metallic phase during treatment. The technology is specifically designed to operate in a hot cell for high or intermediate-level wastes. The small size and internal cold walls of the vessel were chosen for high alpha contamination waste capabilities. The SHIVA process is at TRL 5-6 as a full-scale inactive pilot, which has been tested by the CEA since 1998 for various wastes.

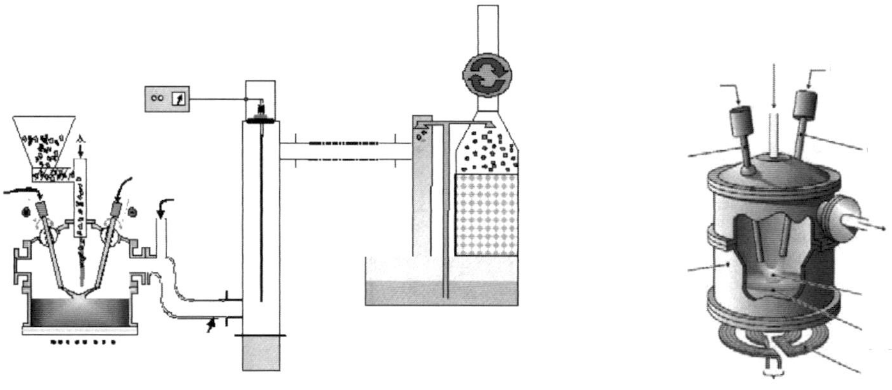

Figure 1. (a) Simplified diagram of the SHIVA process and (b) artist's view of the reactor.

3. Trial information

3.1. Feeds

The waste selected for the trial was a 25 kg mixture of inactive inorganic and organic ion exchange media composed of zeolites, diatoms, strong acid ion exchange resins (IERs) and strong base IERs (Table 1). The zeolites originated from crushed natural zeolite rock composed of 60 ± 5 wt.% chabazite, having a high specific surface area > 100 $m^2 \cdot g^{-1}$, and a cation exchange capacity of 2 $meq \cdot g^{-1}$. Diatomite, being a highly porous material, is usually used as a filtration and clarification medium once cleaned and crushed. The IERs were styrene divinylbenzene copolymers with a strong acid site $-SO_3^- H^+$ for cation exchange (≥ 1.90 $eq \cdot L^{-1}$ H^+ form) and a strong base site ($-N(CH_3)_3^+OH^-$) for anion exchange (≥ 1.20 $eq \cdot L^{-1}$ OH^- form). The IERs contained 50 wt.% of water.

Table 1. Waste feeds for SHIVA trial.

Waste	Fraction (wt.%)	Quantity (kg)	Supplier and reference
Zeolites	45	11.3	SOMEZ Siliz® 14, ZN-024-14
Diatoms	44	11.0	LAFFORT Diatomyl® P0
Strong acid IERs	5.5	1.4	LENNTECH Amberlite™ IRN77
Strong base IERs	5.5	1.4	LENNTECH Amberlite™ IRN78

Vitrification was accomplished by the addition of 40 kg of glass frit whose composition is provided in Table 2. It should be noted that in the first approach, the vitrification was carried out with a known and available glass frit that responds well to induction. Its composition could be optimised for this specific waste as a result of a formulation study.

IOP Conf. Series: Materials Science and Engineering **818** (2020) 012015 doi:10.1088/1757-899X/818/1/012015

Inputs were therefore composed of 38.4 wt.% waste and 61.5 wt.% glass frit. It should be noted that ion exchange media were not exchanged with solutions containing radionuclide surrogates prior to the trial. Therefore, IERs were completely burned. The theoretical composition of the wasteform (Table 2) can be calculated from the theoretical compositions of the glass frit, zeolites, and diatoms, and the amounts of these materials that are introduced into the process.

Table 2. Compositions (expressed in oxide weight percentages) of glass frit, zeolites, diatoms and the wasteform (calculated and analysed by inductively coupled plasma-optical emission spectroscopy (ICP-OES) and energy dispersive X-ray spectroscopy (EDS)). The seven main oxides of the wasteform are written in bold. Note that IERs were completely burned in the process.

Oxide	Glass frit	Zeolites	Diatoms	Wasteform		
				Calculated	Analysed by ICP-OES	Analysed by SEM/EDS
Al_2O_3	5.2	17.0	3.0	**7.2**	**5.6**	**5.7**
B_2O_3	14.7	-	-	**9.8**	**12.1**	n.a.[a]
CaO	4.1	5.5	0.5	**3.8**	**4.2**	**3.9**
CoO	0.7	-	-	0.5	0.6	0.6
Fe_2O_3	4.0	2.5	1.6	**3.4**	**4.2**	**4.0**
K_2O	-	4.5	0.2	0.9	0.5	0.6
Li_2O	2.0	-	-	1.3	1.5	n.a.[a]
MgO	-	1.3	0.3	0.3	0.2	0.4
Na_2O	10.0	-	2.3	**7.1**	**8.7**	**8.0**
Nd_2O_3	7.3	-	-	**4.9**	**6.2**	**6.1**
NiO	0.5	-	-	0.3	0.4	0.3
SiO_2	46.6	51.5	91.1	**57.2**	**50.7**	**52.2**
TiO_2	-	-	0.5	0.1	0.6	0.5
ZnO	2.5	-	-	1.7	2.3	2.1
ZrO_2	2.4	-	-	1.6	2.2	2.1

[a] n.a.: not analysed because boron and lithium are not detected by EDS.

3.2. Heating systems

To ensure complete mineralisation of the waste and guarantee its incorporation into the vitreous matrix, the two modes of heating — induction and plasma torches — were used simultaneously during the waste feeding operations.

- The induction heating system was started at the beginning of the trial and was maintained throughout its duration to maintain a molten glass bath at the bottom of the SHIVA reactor. The induction was started using a titanium starter ring and the generator power was incrementally increased from 40 to 90 kW.
- The sleeves and electrodes of the torches were fed with argon during the start-up phases, and with a mixture of argon and oxygen ($\approx$ 10 $L \cdot min^{-1}$) during the waste introduction periods. The electrodes were pushed a few millimetres at every hour to compensate for their wear.

3.3. Course of the trial

Before the start of the trial, the reactor was pre-filled with 40 kg of glass frit. The waste feed was introduced to the upper part of the reactor; this was achieved during three 8-hour feeds by using a feeding hopper and a worm screw at an average feed rate of 1 $kg \cdot h^{-1}$. This sequencing was used to facilitate a daytime operation for the waste incineration.

For this demonstration, no maximum was sought in terms of processing capacity and waste loading in the vitreous matrix. These parameters would be higher for industrial scale processing.

4. Characterisation of the produced wasteform

The end product of the process was a glass that is macroscopically homogeneous. To be characterised, the wasteform was broken into pieces and a sample was selected by quarter splits (Figure 2).

Figure 2. Wasteform sample (≈ 10 cm long) from the SHIVA trial.

4.1. Microstructure

From a fraction of the Figure 2 sample, a polished cross section was prepared by embedding in an epoxy resin and polishing with diamond abrasive cloths and suspensions before coating with a carbon deposit. Scanning electron microscopy (SEM) (Zeiss Supra™ 55, acceleration voltage of 15 kV) shows a homogeneous material at the micrometric scale (Figure 3 (a)). Another fraction of the sample was crushed and analysed by X-ray diffraction (XRD) (Phillips X'PERT Pro, Bragg-Brentano θ-2θ diffractometer, Cu-Kα_1 radiation, 40 mA-40 kV). The pattern acquired between 10° and 90° (2θ) shows no diffraction peak characteristic of the presence of crystalline phases and has a diffuse scattering characteristic of vitreous materials (Figure 3 (b)). The matrix is thus vitreous and amorphous.

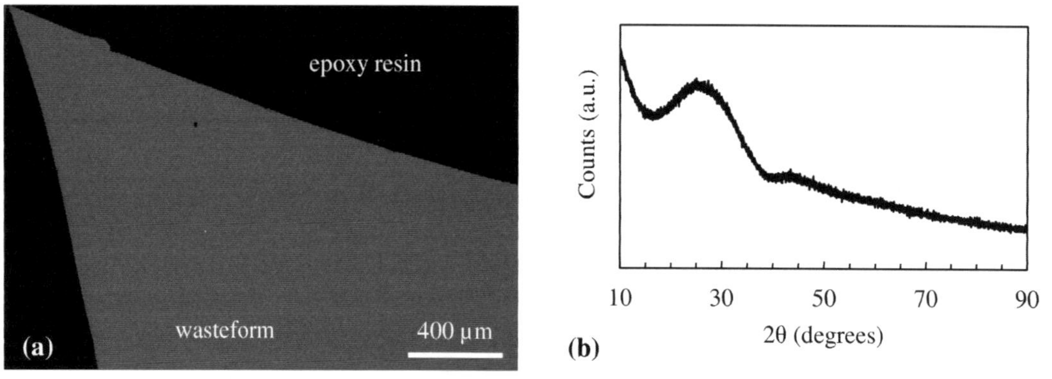

Figure 3. (a) SEM image and (b) XRD pattern of the wasteform.

4.2. Chemical composition

The calculated theoretical chemical composition of the wasteform, shown in Table 2, is checked by two methods: (*i*) complete dissolution of the material followed by solution analysis by ICP-OES and (*ii*) analysis by EDS coupled with SEM. Using EDS, six regions of interest were analysed with similar results (standard deviation ≤ 0.3). The values presented in Table 2 are the average of the six measurements. Since boron and lithium are too light to be analysed by this method, the composition is normalised using the theoretical contents of these two elements.

The analysis of the glass composition by the two techniques provides consistent results. The glass contains mainly SiO_2, B_2O_3, Al_2O_3, Na_2O, and Nd_2O_3 that represent ≈ 85 wt.% of the glass composition. These contents are of the same order of magnitude as those of the inactive surrogate of the French reference high level nuclear wasteglass R7T7. Differences are noted between the theoretical and analysed compositions. They can be explained by an inhomogeneous incorporation of the waste into the matrix or some carryover of the powdery waste in the off-gas treatment system.

4.3. Chemical durability

The chemical durability of the material was studied according to a protocol adapted from the PCT-B standardised test [1]. After crushing the material, 125–250 μm powder was placed in contact with pure water with a glass-surface-area-to-solution-volume ratio (*S/V*) of 10 m^{-1} at 90±2 °C in unstirred perfluoroalkoxy (PFA) reactors. The powder specific surface area of 280 $g \cdot cm^{-2}$ was estimated by adsorption of krypton on the sample surface (Micromeritics ASAP 2020) according to the Brunauer–Emmett–Teller theory [2]. Solution samples, taken at regular intervals, were filtered with a cut-off of 0.45 μm, acidified with ultrapure grade HNO_3 and analysed by ICP-OES (Thermo Scientific iCAP™ 6000 Series). The concentrations were used to calculate normalised mass losses $\mathrm{NL}_i = C_i/(x_i \times S/V)$, where C_i is the concentration of the element *i*, and x_i is the mass percentage of *i* in the glass (using the analysed sample composition: $x_{\mathrm{Al}} = 0.030, x_{\mathrm{B}} = 0.038, x_{\mathrm{Ca}} = 0.030, x_{\mathrm{Li}} = 0.007, x_{\mathrm{Na}} = 0.065, x_{\mathrm{Si}} = 0.237$). The alteration rate, $r = \mathrm{dNL}_i/\mathrm{d}t$, was calculated by linear regression.

The stoichiometric dissolution of the glass during the first days (Figure 4 (a)) gradually becomes non-stoichiometric with a retention of Ca and Al of 4% and 30%, respectively, (relative to B) after one month. The rather unusual shape of the curves $\mathrm{NL}_i = f(t)$ — in particular, B release is not showing a drastic rate drop — is explained by the conjunction of two phenomena: (*i*) the effect of the pH variation on the rate of hydrolysis and (*ii*) the effect of the formation of a surface layer on the alteration rate. The pH increases from 7 to 8.5, which is the value at which it stabilises. This variation causes the increase in the rate of hydrolysis (v_0) of the vitreous matrix. This effect of pH is illustrated by the grey dashed curve on Figure 4 (a), calculated according to: $v_0 = k_+ \times C_{\mathrm{H}^+}^n \times \exp[-E_{\mathrm{a}}/(R \times T)]$, where k_+ is estimated from the $\mathrm{NL_B}$ value at day 1 (0.12 $g \cdot m^{-2}$), *n* is the pH dependence coefficient, E_a is the apparent activation energy, *R* is the ideal gas constant, and *T* is the temperature. In this calculation, $n = 0.4$ and $E_a = 76$ $kJ \cdot mol^{-1}$ are assumed to be equal to those of the R7T7 glass. The progressive divergence between the calculated and the experimental data is explained by the decrease of the alteration rate with the formation of a surface layer.

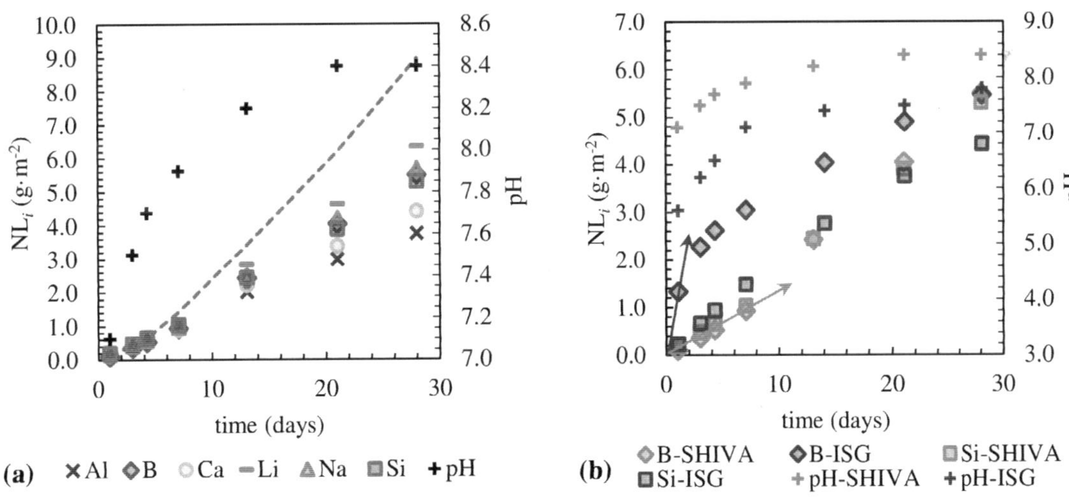

Figure 4. (a) Evolution of NL and pH during wasteform leaching. (b) Compared evolutions of NL and pH during the leaching of ISG and SHIVA-produced wasteform. The arrows are a guide for the eyes to estimate the minimum hydrolysis rate of the matrices.

Comparison of these results with those acquired for the reference International Simple Glass (ISG) [3, 4] altered under the same conditions (Figure 4 (b)) shows a similar (and classical) evolution of the pH of the solutions, with release of alkali species leading to an increase of solution pH. The pH stabilizes at values close to 8 at 90 °C under these experimental conditions of low *S*/*V* ratio. The hydrolysis rate

IOP Conf. Series: Materials Science and Engineering **818** (2020) 012015 doi:10.1088/1757-899X/818/1/012015

of the ISG is higher by approximately one order of magnitude than the rate from SHIVA-produced sample. The dissolution of the ISG is prematurely non-stoichiometric and coincides with the formation of an alteration layer. Due to the diffusion of elements through this alteration layer, the alteration rate of ISG drops earlier than for the SHIVA-produced wasteform. Drawing conclusions about the long-term behaviour of this wasteform would require further investigation.

5. Conclusions

The SHIVA trial conducted in the framework of the THERAMIN project demonstrated the success of the process for the thermal treatment of a mixture of organic and mineral waste composed of zeolites, diatoms and ion exchange resins. The waste loading of 38 wt.% is high and could be increased in the future. The activity of maximising the waste load and the processing capacity was not attempted during this feasibility trial. The final product consists of an amorphous glass, which is primarily composed of SiO_2, B_2O_3, Al_2O_3, Na_2O, and Nd_2O_3. Differences are observed between the theoretical composition of the glass and the composition analysed by two different techniques; this is probably due to low convection in the melter and some carryover of the powdery materials in the off-gas treatment system.

Leaching tests conducted at 90 °C and a low glass-surface-area-to-solution-volume ratio show that the hydrolysis rate of SHIVA-produced wasteform is significantly lower than that of the reference International Simple Glass (ISG). Drawing conclusions about the long-term behaviour of the wasteform would require further investigation, but these results can be regarded as encouraging. Finally, the vitreous nature of the matrix and the presence of boron, which is known to be a tracer for the alteration [5], render it possible to apply proven methodologies to this matrix for elucidating its chemical durability and long-term behaviour.

Acknowledgements

This project has received funding from the Euratom research and training programme 2014–2018 under grant agreement No. 755480. This paper reflects only the authors' views, and the European Commission is not responsible for any use that may be made of it.

The authors are grateful to Aldo Russello for his technical assistance during the trial, Thierry Blisson for the sample preparations, Céline Marcou for the leaching tests, Charlène Vallat for the SEM/EDS analyses, Géraldine Parisot for the ICP-OES analyses, and Myriam Chartier for the XRD analyses.

References

[1] ASTM International 2014 *C1285-14, Standard Test Method for Determining Chemical Durability of Nuclear, Hazardous, and Mixed Waste Glasses and Multiphase Glass Ceramics: The Product Consistency Test (PCT)*

[2] Brunauer S, Emmett P H and Teller E 1938 Adsorption of gases in multimolecular layers *J. Am. Chem. Soc.* **60** 309–19

[3] Gin S, Abdelouas A, Criscenti L J, Ebert W L, Ferrand K, Geisler T, Harrison M T, Inagaki Y, Mitsui S, Mueller K T, Marra J C, Pantano C G, Pierce E M, Ryan J V, Schofield J M, Steefel C I and Vienna J D 2013 An international initiative on long-term behavior of high-level nuclear waste glass *Mater. Today* **16** 243–8

[4] Kaspar T C, Ryan J V, Pantano C G, Rice J, Trivelpiece C, Hyatt N C, Corkhill C L, Mann C, Hand R J, Kirkham M A, Crawford C L, Jantzen C M, Du J, Lu X, Harrison M T, Cushman C, Linford M R and Smith N J 2019 Physical and optical properties of the International Simple Glass *npj Mater. Degrad.* **3** 15

[5] Jantzen C M, Brown K G and Pickett J B 2010 Durable glass for thousands of years *Int. J. Appl. Glass Sci.* **1** 38–62

IOP Conf. Series: Materials Science and Engineering **818** (2020) 012016 doi:10.1088/1757-899X/818/1/012016

Dissolution of plasma treated non-radioactive surrogate cemented concentrates and ion exchange resins in KOH solution at 40 °C

K Ferrand[1], S Caes[1], K Lemmens[1] and K Meert[2]

[1]R&D Waste Packages, Institute for Environment, Health and Safety, Belgian Nuclear Research Centre, Boeretang 200, 2400 Mol, Belgium
[2]ONDRAF/NIRAS, Kunstlaan 14, 1210 Brussel, Belgium

karine.ferrand@sckcen.be

Abstract. In Belgium, plasma incineration is a possible option to (re-)condition low and intermediate level short-lived waste streams in order to obtain end products which can fulfill the criteria developed for a near surface disposal facility. In this study, the chemical durability of plasma treated non-radioactive surrogate cemented concentrates and ion exchange resins (or plasma slags) was investigated by performing MCC-1 leach tests in a KOH solution at pH 13.5 and 40 °C. Results showed that initial dissolution rates of plasma slags were similar to that of SON68-I glass, and final dissolution rates around one order of magnitude lower. SEM analysis revealed the presence of alteration layers rich in K, Ca and Fe, which probably act as diffusion barriers and, associated with the increase of Si and Al concentration in solution, lead to the decrease of the plasma slag dissolution rates.

1. Introduction

Thermal techniques are frequently used for the treatment or conditioning of radioactive waste as significant volume and hazard reductions can be obtained. Vitrification is the most commonly used thermal treatment for reprocessing solutions, but the conventional vitrification installations are not suitable for all waste streams. For these more problematic waste streams, worldwide, different alternative thermal techniques are under development. A problematic waste stream particularly relevant for Belgium concerns cemented low and intermediate level short lived waste showing alkali silica gel formation. A few years ago, the formation of such alkali silica gels was observed on cemented concentrates and on some drums containing cemented resins from nuclear installations [1]. Gel formation might compromise the stability of the immobilization matrix and its swelling could affect the mechanical integrity of the engineered barrier system of the disposal. Therefore, a R&D programme was developed which defines four scenarios for the management of these drums. One of these scenarios considers the full reconditioning by a thermal treatment. Plasma incineration was selected, because it is believed to be a versatile technique for the (re-) conditioning of both unconditioned and conditioned waste streams. In the past, such a technique was successfully used to vitrify PuO_2 simulant and CeO_2 in an amorphous calcium iron aluminosilicate component of a product slag with simultaneous destruction of the organic and polymer waste fractions [2]. The end products should be disposable, this means that their compliance with the conformity criteria developed for the disposal system (near surface disposal) should be proven. One of these waste acceptance criteria is chemical durability. In the

IOP Conf. Series: Materials Science and Engineering **818** (2020) 012016 doi:10.1088/1757-899X/818/1/012016

framework of the EC THERAMIN project, this chemical durability was tested on a series of test samples, to demonstrate the potential of the technique for this type of waste. As samples, surrogate non-radioactive concentrates and resins were made, cemented and reprocessed using the plasma torch technique. The methodology for the preparation of such surrogate waste forms, described in detail in [3], was based on assumptions on the plasma torch operational parameters and the available information on the cementation campaigns for the real waste. In this paper, we investigate the dissolution of six different plasma treated non-radioactive surrogate waste forms in a KOH solution at pH 13.5 and 40 °C in order to assess their chemical durability in disposal conditions dominated by the presence of high pH cement.

2. Plasma treated surrogate waste forms

Surrogate non-radioactive concentrates and resins were made, cemented and reprocessed at the Phoenix Solutions Hutchinson Test Facility using a plasma torch power of 1200 kW. The plasma treated surrogate waste forms, or plasma slags, originating from concentrates are referred to as Simuli samples (from 1 to 4) and those from resins as R1 IRN 77 and R2 IRN 78. Their composition, determined by X-Ray Fluorescence (XRF), is indicated in Table 1. Plasma slags were composed of 38 – 50 wt % SiO_2, 12 – 30 wt % Fe_2O_3, 21 – 28 wt% CaO and 5 – 15 wt% Al_2O_3. For some plasma slags, low contents in sodium, magnesium and zinc oxides were also detected (< 2.2 wt%).

Table 1. Plasma slags composition (in wt. %) from XRF.

	SiO_2	Fe_2O_3	CaO	Al_2O_3	NaO	MgO	ZnO
Simuli 1	41.1	21.9	23.6	7.6	1.9		
Simuli 2	40.1	30.5	20.7	5.0			1.5
Simuli 3	49.5	14.0	25.6	6.0	2.1		1.1
Simuli 4	50.0	15.4	25.1	6.5			1.0
R1 IRN 77	38.2	19.0	22.7	14.7		1.7	1.0
R2 IRN 78	45.6	11.8	27.9	9.0		2.2	

Scanning Electron Microscopy (SEM) and elemental mappings showed that Simuli 1, Simuli 3, Simuli 4 and R2 IRN 78 samples were amorphous, contrary to Simuli 2 and R1 IRN 77 in which crystalline phases, i.e. bright and light grey dendrites, were observed (Figure 1). By combining Energy Dispersive X-ray (EDX) and X-Ray Diffraction (XRD) analysis these phases were identified as magnetite (Fe_3O_4) and clinopyroxene ($CaAl_{0.7}Fe_{0.2}Mg_{0.6}Si_{1.5}O_6$). The fractions of amorphous and crystalline phases were quantified by mixing the milled samples with about 9% ZnO, which served as internal reference. The Simuli 2 and R1 IRN 77 samples were partially amorphous (48.7% and 42.7%, respectively) and contained magnetite (13.9% and 2.1%, respectively) and clinopyroxene (36.7% and 48.3%, respectively). For R1 IRN 77 other diffraction peaks that might be attributed to plagioclase (6.6%) were also detected. SEM images also evidenced the presence of many pores, with a size up to 50 µm, in these two plasma slags.

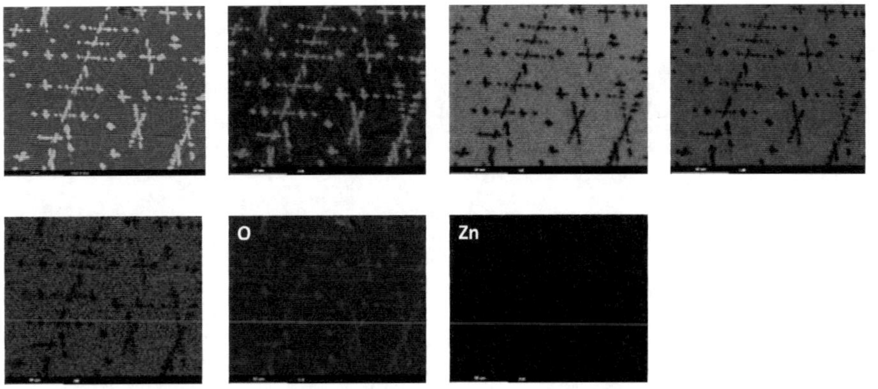

Figure 1. SEM image and elemental mappings for the Simuli 2 sample.

3. Dissolution methodology

Plasma slag dissolution experiments based on the MCC-1 static Leach Test Method [4] were carried out in a KOH solution at $pH_{(25\,°C)}$ 13.5 ± 0.2 and at 40 ± 0.1 °C. In a glove box under an argon atmosphere the plasma slags, cut into rectangular coupons of 5 x 10 x 2 mm (width x length x thickness) and inserted into a Teflon® sample holder in a vertical position, were placed in a 25 mL Teflon® container. Then, 17 mL of the leaching solution was added to reach the low sample surface area to solution volume ratio of 10 m-1, which guarantees that the concentration of rate controlling elements remains below the saturation concentration, and that the dissolution rate remains relatively high, allowing the relative durability of the different slags to be distinguished. Similar tests were performed with the French inactive reference nuclear glass SON68, but without U and Th in its composition (denoted as SON68-I), which is thus used as a reference material. At the end of each leaching test, i.e. after 1, 2, 4, 7, 14, 28, 56 and 182 days, 5 mL of solution was collected using a syringe and filtered on a 0.20 μm membrane. An aliquot of solution was used for pH measurement, and 1 mL was diluted with 2 mL of ultrapure water, ultrafiltered at 6000 RPM for 20 minutes, and diluted with 2% ultrapure HNO_3 to determine the elemental concentrations by ICP-AES (Thermoscientific Iris Intrepid II dualview). The normalized mass loss (NL in g m^{-2}) from the Si and Al released from the sample to the solution was calculated according to Equation (1):

$$\boldsymbol{NL(i)} = \frac{(C_i . V)}{(f_i\, SA)} \tag{1}$$

Where C_i is the concentration of the element i in the aliquot of solution in g m^{-3} (or mg L^{-1}), V is the total volume of solution in m^3, f_i is the weight percentage of element i in the unaltered sample, SA is the total surface area of the exposed sample in m^2.

Tests with a duration of 28 days were done in triplicate, and blank tests, i.e. containers with only the leaching solution, were also carried out.

4. Alteration layer analysis methodology

The plasma slags altered for 182 days were removed from the solution, rinsed with ultrapure water, left to dry at room temperature in the glove box prior to be analyzed by XRD, and then embedded in epoxy resin for SEM analysis.

Plasma slags were scanned using a Bruker D8 advance apparatus with the following settings: spinning (15 min^{-1}), radial soller slit 2.5°, automatic divergence slit, 2θ range 10 – 90°, step size 0.03°, CuKα LFF X-ray tube operated at 40 kV and 40 mA, and a position-sensitive X´Celerator detector. The resulting XRD patterns were evaluated with the aid of the X´Pert High Score Plus software and the PDF-2 Mineral Database.

SEM analysis of the cross sections coated with a 3 nm C layer was performed on a JEOL JSM 6610 with an accelerating voltage of 15 keV and a work distance of 10 mm. The microscope was coupled with an Energy Dispersive Electron Spectroscopy detector providing semi-quantitative analysis.

5. Results and discussion

5.1. Solution analysis

The mean pH of the aliquots of solution collected over the test period, measured at room temperature, was approximatively pH 13.5; it was similar to the mean pH of the blank solutions indicating that the plasma slag dissolution did not significantly affect the pH of the system.

Figure 2 shows the Si, Al and Ca concentrations as a function of time for the Simuli 2, R1 IRN 77 and SON68-I samples. The concentrations measured for the other plasma slags, not presented here, were in the same range of values as those of the Simuli 2 and R1 IRN 77 samples. The silicon concentration increased from 1 – 2 mg L^{-1} up to 20 – 31 mg L^{-1} except for the SON68-I sample for which a higher Si concentration of 56 mg L^{-1} was measured. The aluminium concentration increased from 0.4 – 1 mg L^{-1} up to 5 – 18 mg L^{-1}. The calcium concentration increased up to 6 – 13 mg L^{-1}, and then slightly decreased down to 5 – 6 mg L^{-1}. Very low iron concentrations (< 1 mg L^{-1}) were measured and the K concentration was the same as in the blank solutions.

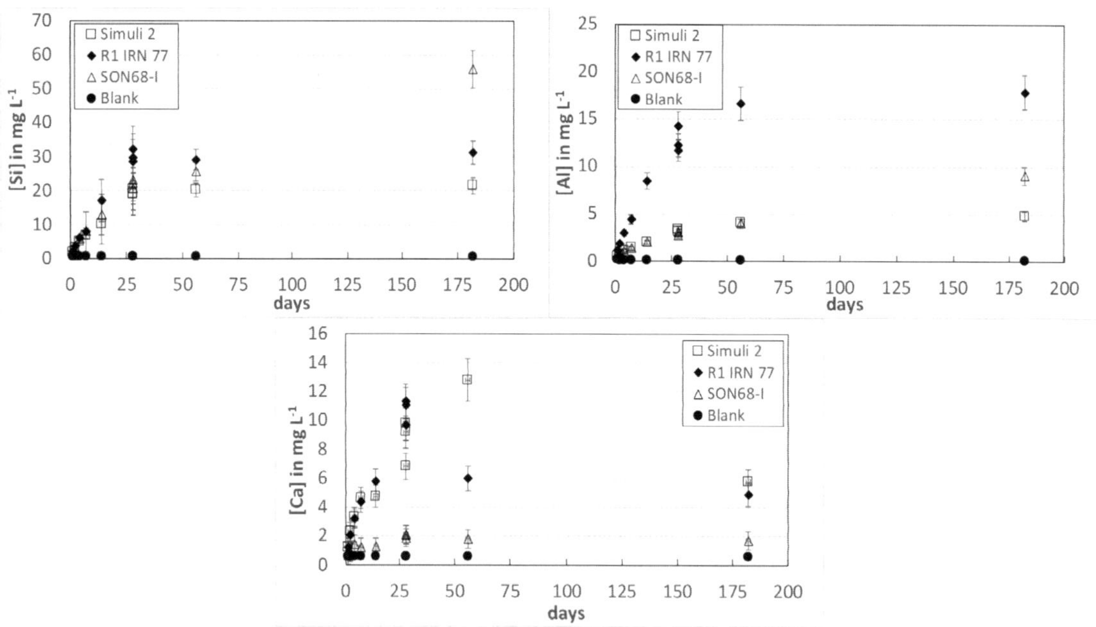

Figure 2. Si, Al and Ca concentrations (mg L^{-1}) as a function of time for the Simuli 2, R1 IRN 77, SON68-I samples and a blank test (leach tests in KOH solution at $pH_{(25\,°C)}$ 13.5 ± 0.2 and at 40 ± 0.1 °C).

5.2. Plasma slag dissolution rates

The Si and Al normalized mass losses as a function of time are presented in Figure 3 with the 95% confidence intervals calculated by error propagation. The slopes of the linear regressions give the sample dissolution rates, with the initial and final dissolution rates determined by connecting the NL values after 1 day with the origin (0,0) and those after 56 and 182 days, respectively. For all the samples, a maximum dissolution rate was observed after 1 day, followed by a dissolution rate drop that could be due to the increase of the concentration of the glass constituents (Si, Al,...) that are involved in the formation of a protective layer whose morphology depends on the type of sample. Such dissolution behaviour is consistent with results reported for many glass compositions altered under various experimental conditions [5 – 7].

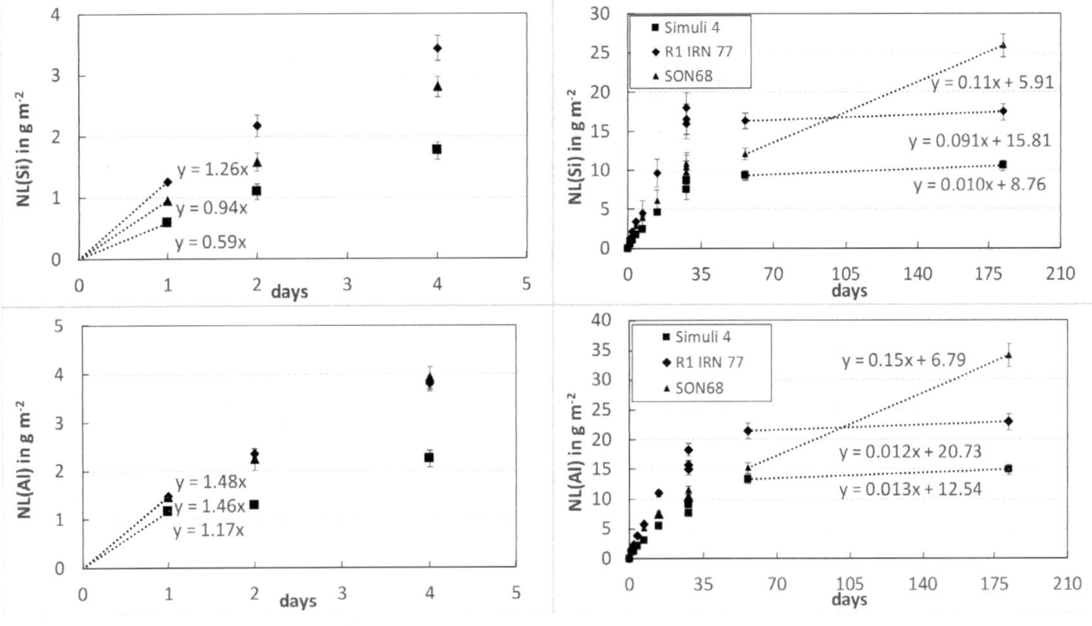

Figure 3. Si and Al normalized mass losses as a function of time for the Simuli 4, R1 IRN 77 and SON68-I samples in leach tests in KOH solution at $pH_{(25\,°C)}$ 13.5 ± 0.2 and at 40 ± 0.1 °C. The graphs on the left hand side show only the data up to 4 days, to focus on the initial dissolution rate; the graphs on the right hand side show all data, to focus on the final dissolution rate.

Table 2 summarizes the initial and final dissolution rates based on NL(Si) and NL(Al) determined for all the samples. In the period 0 – 1 day, the initial dissolution rates based on NL(Si) and NL(Al) were in the range of 0.6 – 1.3 g m^{-2} d^{-1} and 1 – 1.9 g m^{-2} d^{-1}, respectively. Because until 28 days of leaching the uncertainties on the Si concentrations were much higher than those on the Al concentrations, the uncertainties on the dissolution rates calculated from NL(Si) were also much higher meaning that the rates based on NL(Al) were more accurate. Initial dissolution rates were slightly higher for the Simuli 2 and R1 IRN 77 samples, which were the porous plasma slags containing magnetite and clinopyroxene.

The initial dissolution rates based on NL(Si) and NL(Al) for the SON68-I sample were equal to 0.94 g m^{-2} d^{-1} and 1.46 g m^{-2} d^{-1}, respectively. They were similar to those found for the plasma slags suggesting that, under alkaline conditions, the dissolution behaviour of the plasma slags is similar to that of a well performing nuclear glass. The initial dissolution rate based on NL(Al) for the SON68-I sample was close to the maximum dissolution rate of 1.4 g m^{-2} d^{-1} calculated from the equation established by [8] considering the leaching tests data of 19 nuclear glasses at alkaline pH (up to 12). This indicates that the initial dissolution rates established after one day and calculated from the aluminium release can be considered as maximum dissolution rates.

The final dissolution rates of the plasma slags were similar; they were in the range of 0.01 – 0.03 g m^{-2} d^{-1}, except for the R2 IRN 78 sample for which a rate one order of magnitude lower was found (0.004 g m^{-2} d^{-1}). The final plasma slags dissolution rates were around one order of magnitude lower than those determined for the SON68-I sample indicating thus a better chemical durability under alkaline conditions and at 40 °C for the plasma slags. However, long-term experiments have still to be performed to draw conclusions about the long-term chemical durability.

Table 2. Dissolution rates (g m^{-2} d^{-1}) calculated from NL(Si) and NL(Al) for the plasma slags and SON68-I altered in a KOH solution at pH 13.5 and 40 °C.

Samples	Initial dissolution rate (g m^{-2} d^{-1}) in the period 0 – 1 day		Final dissolution rate (g m^{-2} d^{-1}) in the period 56 – 182 days	
	From NL(Si)	From NL(Al)	From NL(Si)	From NL(Al)
Simuli 1	0.66 ± 0.31	1.04 ± 0.12	0.031 ± 0.020	0.055 ± 0.020
Simuli 2	1.07 ± 0.34	1.94 ± 0.22	0.0052 ± 0.010	0.022 ± 0.020
Simuli 3	0.60 ± 0.27	1.23 ± 0.14	0.015 ± 0.010	0.052 ± 0.010[a]
Simuli 4	0.59 ± 0.27	1.17 ± 0.13	0.010 ± 0.010	0.013 ± 0.020
R1 IRN 77	1.26 ± 0.35	1.48 ± 0.17	0.091 ± 0.020	0.012 ± 0.030
R2 IRN 78	0.75 ± 0.29	1.31 ± 0.15	0.0045 ± 0.010	0.011 ± 0.020

[a] Rate between 28 days and 182 days.

5.3. Alteration layer

At the end of the tests, i.e. after 182 days, alteration products were observed on the surface of the plasma slags. For the amorphous samples, i.e. Simuli 1, Simuli 3, Simuli 4 and R2 IRN 78 SEM images showed an altered layer, comprising four sublayers: a dark porous layer close to the unaltered sample, a less porous grey layer, a second dark porous layer and a very thin outer layer (Figure 4 a and b illustrates the alteration layer observed on the Simuli 4 sample). Elemental analysis of the alteration layer using SEM-EDX indicated that in comparison to the unaltered sample, the alteration layer was depleted in Si, Al and Na, and enriched in Ca, Fe and K. However, in the very thin outer layer in which the highest K content was measured, Fe was only present in a very small quantity.

For the Simuli 2 and R1 IRN 77 samples containing magnetite and clinopyroxene a uniform alteration layer depleted in Si, Al and Na, and enriched in Ca, Fe and K with respect to the dark grey amorphous matrix was observed (Figure 4 c and d shows the alteration layer observed on the R1 IRN 77 sample). Combined with the decrease in Ca concentration and very low Fe concentration in solution, SEM analysis confirmed that, whatever the type of plasma slag, Ca and Fe were retained in the alteration layer. The decrease in K concentration was not observed due to the very high initial concentration in the leaching solution, but K retention in the alteration layer was reported in many studies dealing with glass alteration in YCWCa, a synthetic cementitious water rich in K [9, 10].

A mono-layer enriched in Ca, Fe, K and Zn was also present on the surface of the SON68-I sample, but Ca and Fe contents were respectively 2 and 3 – 5 times lower than those measured in the alteration layers of the plasma slags. Thanks to the high Ca and Fe content, the alteration layers (mono- or multi-layers) developed on the surface of the plasma slags were thus more protective than the one formed on the SON68-I sample, leading to lower final dissolution rates for the plasma slags. Ca is indeed known to have a protective effect at high pH and high sample surface area to solution volume [11].

XRD analysis conducted on the surface of the Simuli 2 and R2 IRN 78 samples altered for 182 days did not reveal the presence of crystalline phases other than magnetite and clinopyroxene, which were initially present in the samples. Diffraction peaks that might be attributed to magnetite were detected at the surface of the Simuli 3, Simuli 4 and R2 IRN 78 samples whereas no diffraction peaks were observed for SON68-I.

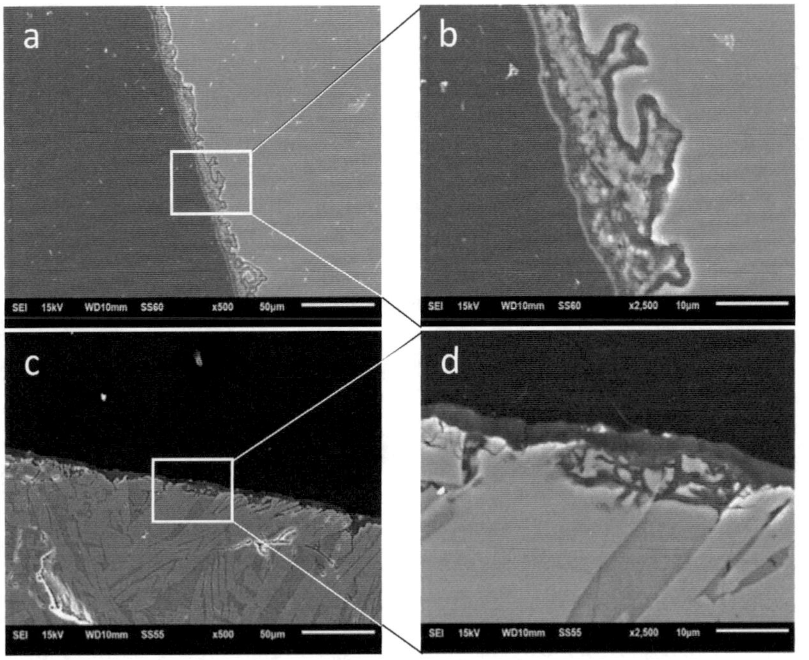

Figure 4. SEM images of the cross sections of the embedded altered samples: a and b) Simuli 4 sample (amorphous plasma slag); c and d) R1 IRN 77 sample (plasma slag containing magnetite and clinopyroxene).

6. Conclusions

The chemical durability of plasma treated non-radioactive surrogate cemented concentrates and ion exchange resins (plasma slags) was investigated by conducting MCC-1 leach tests with a KOH solution at pH 13.5 and at 40 °C. Initial rates calculated from NL(Si) and NL(Al) in the period 0 – 1 day were in the range of 0.6 – 1.3 g m^{-2} d^{-1} and 1 – 1.9 g m^{-2} d^{-1}, and were similar to those of SON68-I glass altered under the same conditions. SEM analysis performed on the samples leached for 182 days showed the presence of a multi-layer for the plasma slags initially amorphous and a mono-layer for those containing crystalline phases. Compared to the unaltered samples, these alteration layers were very rich in Ca, Fe, and also contained K originating from the leaching solution. The formation of such alteration layers combined with the increase of Si and Al concentration in solution leads to a decrease of the dissolution rates by two orders of magnitude. The final plasma slag dissolution rates in the range of 0.01 – 0.03 g m^{-2} d^{-1} were about one order of magnitude lower than those of SON68-I glass, suggesting a better chemical durability. However, this should be confirmed by conducting experiments on a longer term.

Acknowledgements

This project has received funding from the Euratom research and training programme 2014-2018 under Grant Agreement No 755480. The paper reflects only the author's view and the Commission is not responsible for any use that may be made of it. This research was also financed by ONDRAF/NIRAS, the Belgian Agency for the Management of Radioactive Waste and Fissile Materials, as part of the research program on 'B&C wastes'. The authors gratefully acknowledge the technical support received from Pieter Schroeders and Ben Gielen.

References

[1] NIROND 2017 Note 2017-2963 *RD&D program for the potentially ASR-affected Conditioned Waste Packages* 18/12/2017 ONDRAF/NIRAS Brussels, Belgium

[2] Hyatt N, Morgan S, Stennett M, Scales C and Deegan D 2006 Characterisation of Plasma Vitrified Simulant Plutonium Contaminated Material Waste *Mater. Res. Soc. Symp. Proc.* **985** 0985-NN10- 06

[3] NIROND 2017 Note 2017-2475 *Methodology for the formulation of surrogate cemented concentrates for the assessment of the plasma torch technique* 14/08/2018 ONDRAF/NIRAS, Brussels, Belgium

[4] ASTM 2004 *Standard test method for static leaching of monolithic waste forms for disposal of radioactive waste* ASTM C1220-98

[5] Poinssot C and Gin S 2012 Long-term Behavior Science: The cornerstone approach for reliably assessing the long-term performance of nuclear waste *J. Nucl.Mater.* **420** 182–92

[6] Frugier P, Chave T, Gin S and Lartigue J-E 2008 Application of the GRAAL model to leaching experiments with SON68 nuclear glass in initially pure water *J. Nucl. Mater.* **380** 8–21

[7] Van Iseghem P, Aertsens M, Lemmens K, Gin S, Deneele D, Grambow B, McGrail P, Strachan D, Wicks G and McMenamin T 2007 *GLAMOR: A critical evaluation of the dissolution mechanisms of high-level waste glasses in conditions of relevance for geological disposal* Final report on ER contract No. FIKW-CT-2001-20140

[8] Vienna J, Neeway J, Ryan J and Kerisit S 2018 Impacts of glass composition, pH, and temperature on glass forward dissolution rate *npj Materials degradation* **22** 1–22

[9] Ferrand K, Liu S and Lemmens K 2013 The interaction between nuclear waste glass and ordinary portland cement *Int. J. Appl. Glas. Sci.* **4** 328–40

[10] Mann C, Ferrand K, Liu S, Eskelsen J R, Pierce E, Lemmens K and Corkhill 2019 Influence of young cement water on the corrosion of the International Simple Glass *npj Mater. Degrad.* **3** 1–9

[11] Mercado-Depierre S, Angeli F, Frizon F, Gin S 2013 Antagonist effects of calcium on borosilicate glass alteration *J. Nucl. Mater.* **441** 402–10

IOP Conf. Series: Materials Science and Engineering **818** (2020) 012017 doi:10.1088/1757-899X/818/1/012017

Geopolymerisation of gasified ion-exchange resins, mechanical properties and short-term leaching studies

Tapio Vehmas[1], Emmi Myllykylä[1], Matti Nieminen[1], Jaana Laatikainen-Luntama[1], Markku Leivo[1] and Markus Olin[1]

[1] Technical Research Centre of Finland VTT Ltd. P.O. Box 1000 FI-02044 VTT, Finland

Email: tapio.vehmas@vtt.fi

Abstract. A combination of gasification of low and intermediate level radioactive waste (LILW) and conditioning of the resulting product within a geopolymer matrix is a potential alternative for vitrification technologies as an immobilisation method. Geopolymer matrices have demonstrated good retention capability for radionuclides in many studies and the technology has been implemented at an industrial scale in Slovakia and Czech Republic. However, the practical waste loading has been limited by the mechanical properties of the encapsulated matrix. Even a small amount of ion exchange resin (IXR) decreases the strength of the matrix and cohesion of the matrix is lost when the fraction of resin exceeds 15-20%. In this study, the potential to combine gasification as thermal treatment and various inorganic binders as encapsulation matrices was evaluated. After gasification, the mechanical properties were not similarly sensitive to the encapsulation of IXR. Gasification enabled substantially higher loading of IXR into the sample. Also, gasification of the IXR decreased matrix apparent Caesium diffusion. Very low apparent diffusion coefficients of Cs were calculated when gasified resin was encapsulated in metakaolin matrix. Theoretically, the amount of Cs within the same volume of encapsulated material could be increased by 800 times following gasification and encapsulation within an alkali-activated metakaolin (MK) binder.

1. Introduction

Nuclear wastes are traditionally classified into three categories depending on the material's radioactivity. High level radioactive wastes (HLW) consists of spent nuclear fuel and other high activity products. In Finland, low and intermediate level radioactive wastes (LILW) are items and materials that are contaminated or activated by the operation of a nuclear power plant. LILW contains only 1% of the total radioactive inventory produced by the nuclear industry. However, it represents more than 95% of the total volume of radioactive waste. Partly, due to this high volume, issues related to final disposal of HLW and LILW are very different.

HLW must be isolated from the biosphere for hundred thousand years or even longer. For LILW, the timespan is only from hundreds to thousand years. Relatively small volume of HLW enables use of expensive encapsulation materials (e.g. copper canisters) and deep underground disposal (400-500 meters below ground surface) [1]. With LILW, encapsulation with a binder is usually sufficient and the repositories can be at shallower depth than the HLW repositories.

Binders for LILW encapsulation should stabilize wastes and retard dissolution of radionuclides. Binders from the construction industry are often used due to high availability and low price (e.g. Portland cement and bitumen). However, both have their weaknesses. Bitumen is a combustible material and potential fire could cause release of radionuclides into the environment. Cementitious materials are not combustible but they are vulnerable to various chemical alterations. Small amounts of dissolving organics can prevent hardening process of a cement and even some elements have similar effect (e.g. boron). Hardened Portland cement has also poor adhesion to solid organic materials, which decreases the strength of the cementitious matrix in encapsulation.

Because of the above described reasons, the amount of encapsulated ion-exchange resins (IXR) that can be incorporated in the cementitious matrix is only 10-15% in industrial scale encapsulation. With larger amounts, cohesion of the encapsulation matrix is lost. As the IXR forms a major part of the Finnish ILW inventory, most of the repository space is occupied by supplementary materials, taking space from the actual radioactive waste.

Thermal treatments of IXR are able to reduce the volume of the radioactive waste significantly. Due to removal of organic content in IXR, the volume of the thermally treated material reduces to 1-5% of the original volume [2]. However, remaining radioactive ashes must be encapsulated. Vitrification, encapsulation simultaneously into a glass phase has already been studied.[3]

Another option is to perform thermal treatment and encapsulation sequentially. Gasification is a thermal treatment that can remove organic matter from IXR without release of radionuclide quantities. Binders that are traditionally used for encapsulation are potential option to bind the gasified resins. However, many studies present that alternative inorganic binders have higher radionuclide bounding capacity. Specially, metakaolin (MK) has been used in industrial scale encapsulation in Czech Republic and Slovakia [4]. Increased binding capacity of the encapsulation matrix might become a necessary in gasified resins, as the radionuclide content in the encapsulation increases significantly.

In this study, the potential to combine gasification as thermal treatment and various inorganic binders as encapsulation matrices was tested. Evaluation was performed to Ordinary Portland Cement (OPC), alkali-activated blast furnace slag (BFS) and alkali-activated metakaolin (MK). Results were compared to traditional pristine resin/OPC encapsulation and existing guidelines. Evaluation was performed using inactive cesium tracer as cesium (^{137}Cs, ^{134}Cs) is the main source of radioactivity in IXR within the first 200-300 years.

2. Materials and Methods

Organic IXR was doped with cesium tracer (^{133}Cs). After the doping, the IXR was gasified at 850 °C, steady-state measuring period being 5 hours. Cesium content of the pristine resin was 3.8 mg/kg, and gasified resins Cs content was 70.7 mg/kg. Cs content were measured by dissolving samples and determining the Cs concentration with sector field Inductively Coupled Plasma mass-spectrometer (SF-ICP-MS) (Element 2™ ThermoScientific).

Prior to analysis of cesium, approximately 50 m of resin sample was weighed and solution was added into Teflon vials. A mixture of concentrated 4 ml HNO_3 (suprapure, s.p) and 0.5 ml H_2O_2 was used for resin samples and 2 ml HNO_3 (s.p), 1.5 ml HF and 2 ml $HClO_4$for ash samples. The microwave assisted digestion was run with SRC (single reaction chamber) Ultrawave system by Milestone. Chamber was pressurized to 40 bar and temperature was ramped up in steps 1) 10 min to 60°C, 2) 10 min to 200°C, 3) 5 min to 230°C, and 4) finally kept 10 min at 230°C before cooling down.

After gasification, ashes were homogenized and mixed with studied binder. Ash dosing was calculated as a pristine resin and is presented in Table 1. Studied binders were Ordinary Portland cement CEMI (Annläggninscement from Degerhamn, Cementa Ab), blast furnace slag (Masuunikuonajauhe KJ100, Finnsementti Oy) and metakaolin (Metastar from BASF). Alkaline activators and water was premixed. Potassium hydroxide (Sigma Aldrich, reagent grade) and sodium silicate solution (Zeopol 33 from Huber Engineered Materials) were used as activators. Solution was left to cool down to room temperature prior to sample batching. The resulting solution was combined with solid materials and the mix was agitated for three minutes. Samples were cast into moulds and BFS and MK samples were cured for 24 hours in 80°C at relative humidity 100%. OPC samples were cured in autogeneous conditions at room temperature for 24 hours. Samples were demoulded and cured in autogeneous conditions in room temperature for an additional 6 days. Mix designs of the studied samples are presented in Table 1.

Table 1. Mix designs of the studied samples in measurement of the leachability of solidified low-level radioactive wastes by a short-term test procedure.

Binder system		Dosing /%	Binder /g	Pristine resin /g	Gasified resin /g	Water /g	Sodium silicate[a] /g	Potassium hydroxide /g
OPC	doped resin	13	1000	200	-	400	-	-
OPC	gasified resin	10	1000	-	8.5	400	-	-
BFS	gasified resin	10	1000	-	8.2	105	178.5	119
MK	gasified resin	10	1000	-	14.1	-	1177	202
OPC	gasified resin	20	1000	-	18.7	400	-	-
BFS	gasified resin	20	1000	-	18.8	105	178.5	119
MK	gasified resin	20	1000	-	31.6	-	1177	202
OPC	gasified resin	50	1000	-	74.5	400	-	-
BFS	gasified resin	50	1000	-	74.6	105	178.5	119
MK	gasified resin	50	1000	-	126.6	-	1177	202

[a] Solid content of sodium silicate was 37%.

Leachability of the Cs tracer from the samples was studied according to American Nuclear Society standard: *Measurement of the leachability of solidified low-level radioactive wastes by a short-term test procedure* [5]. In the tests, samples with cylinder geometry were exposed to ion-exchanged water for a certain time periods. Water content was tenfold respect to the surface area of the sample. The concentration of leached tracer was measured and based on the results, Cs apparent diffusion coefficient was calculated. Cs apparent diffusion coefficients were calculated according to Equation 1. Detailed description of the method is available elsewhere.[4] The amount of leached Cs tracer was determined using SF-ICP-MS.

$$D = \pi \left[\frac{a_n/A_0}{(\Delta t)_n}\right]^2 \left[\frac{V}{S}\right]^2 \left[\frac{1}{2}\left(t_n^{1/2} + t_{n-1}^{1/2}\right)\right]^2 \quad (1)$$

In Equation 1, D is the apparent diffusion coefficient (cm^2/s), a_n is the quantity of a tracer released from the specimen (mol), A_0 is the total quantity of a tracer (mol), t is the time (s), Δt in the leaching interval (s), V is the volume of the specimen (cm^3) and S is the geometric surface area of the specimen as calculated from measured dimensions (cm^2).

The influence of the pristine IXR and gasified resin to mechanical properties of OPC matrix was studied. Mix designs reported in Table 2 were cast and the compression strength of the samples were determined at the age of 7 days according to [6].

Table 2. Mix designs of studied samples in compression strength studies.

Binder system	Resin/binder content /%	Binder /g	Pristine resin /g	Gasified resin /g	Water /g
OPC pristine resin	1	1000	14.1	-	400
OPC pristine resin	5	1000	73.7	-	400
OPC pristine resin	10	1000	155.6	-	400
OPC pristine resin	15	1000	247.1	-	400
OPC gasified resin	1	1000	-	14.1	400
OPC gasified resin	5	1000	-	73.7	400
OPC gasified resin	10	1000	-	155.6	400
OPC gasified resin	15	1000	-	247.1	400

3. Results

Cesium concentrations from short-term leaching test procedure are presented in Table 3. Dosing of resin is presented as pristine resin, calculated from cesium content of pristine resin and gasified resin. Gasification of the IXR changed the resin's behaviour in the leaching test. The amount of leached cesium decreased due to gasification treatment of the resin. Concentration of the leached cesium was approximately ¼ in OPC/gasified resin samples compared to OPC/pristine resin samples. Even greater difference was observed when the binder was changed to BFS or MK. With BFS/ gasified resin samples, the concentration of leached cesium was approximately 1/10 compared to OPC/pristine resin samples. It should be noted that during sequential leaching, leached cesium concentrations decreased in OPC-based samples whereas concentrations in BFS-based samples tend to increase. Cesium concentrations leached from MK/gasified resin samples were very low and even below detection limit in some cases. Even with the 50% dosing, leached cesium concentrations were very low from the MK/gasified resin combination.

Calculated apparent Cs diffusion coefficients are presented in Figure 1. Average apparent diffusion coefficient was 1.1×10^{-7} cm^2/s for OPC/doped resin combination. For OPC/gasified resin, the average apparent diffusion coefficient was 6.2×10^{-9} cm^2/s. Average apparent diffusion coefficients for BFS/gasified resin and MK/gasified resins were 1.0×10^{-9} cm^2/s and 1.1×10^{-12} cm^2/s, respectively. Apparent diffusion coefficients enabled calculations related to maximum encapsulated cesium contents related to released cesium content. Assuming identical cases (exposure times, geometry and leachate volumes), the dosing of initially encapsulated material becomes proportional to apparent diffusion coefficients according to Equation 2.

$$P_{eq} = \frac{A_0}{A_0^{ref}} = \sqrt{\frac{D^{ref}}{D}} \tag{2}$$

In Equation 2, P_{eq} is the proportional encapsulated resin content for equal leachate concentrations, A_0^{ref} is the total quantity of the tracer in reference sample and D^{ref} is the apparent diffusion coefficient of reference sample (cm^2/s). Table 4 presents the calculated proportional encapsulated resin contents for equal cesium leachate concentrations. On average, gasification of the resin enabled five times more cesium to be included within the OPC matrix. With BFS and MK -binders, gasification of the resin enabled 9 and 800 times more cesium in the samples.

Table 3. Measured leachate concentrations from short-term leaching test procedure. Dosing corresponds to the amount of pristine resin.

Binder system		Dosing /%	Leachate concentrations (ppb)					
			2h	7h	24h	48h	72h	96h
OPC	doped resin	13	4.53	4.07	4.65	4.32	2.35	1.31
OPC	gasified resin	10	1.53	1.10	1.38	1.60	1.35	1.01
BFS	gasified resin	10	0.32	0.29	1.06	1.00	0.73	0.67
MK	gasified resin	10	0.01	0.00	0.02	[a]	[a]	[a]
OPC	gasified resin	20	1.41	0.97	1.13	1.11	0.93	0.67
BFS	gasified resin	20	0.72	0.57	1.92	1.89	1.47	1.17
MK	gasified resin	20	0.00	0.01	0.02	[a]	[a]	[a]
OPC	gasified resin	50	4.26	3.43	5.53	5.97	5.60	4.94
BFS	gasified resin	50	1.43	1.86	4.40	5.24	3.98	3.25
MK	gasified resin	50	0.03	0.04	0.06	0.04	0.01	0.00

[a] Under detection limit.

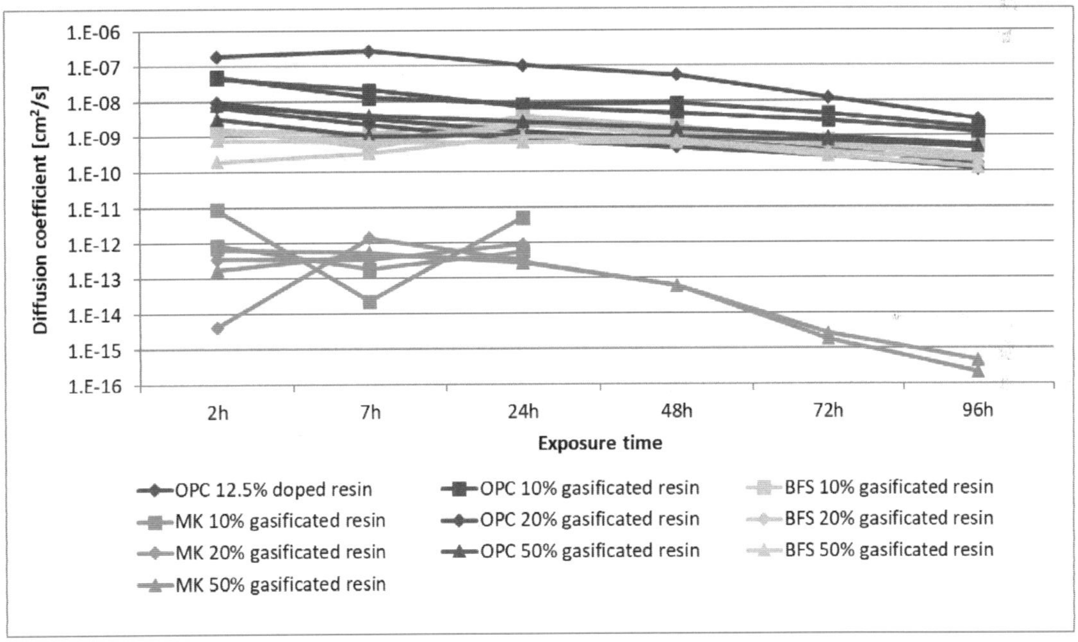

Figure 1. Calculated apparent Cs diffusion coefficients from short-term leaching test procedure.

Table 4. Calculated proportional initial cesium contents for equal cesium leachate concentrations.

Binder system		Dosing /%	Proportional encapsulated resin contents (P_{eq})					
			2h	7h	24h	48h	72h	96h
OPC	doped resin	13	1	1	1	1	1	1
OPC	gasified resin	10	2	3	4	3	2	1
BFS	gasified resin	10	11	15	6	5	4	2
MK	gasified resin	10	198	1361	188	-	-	-
OPC	gasified resin	20	5	8	10	9	6	5
BFS	gasified resin	20	11	17	7	7	5	3
MK	gasified resin	20	1036	469	406	-	-	-
OPC	gasified resin	50	6	9	7	6	4	2
BFS	gasified resin	50	20	19	11	9	6	4
MK	gasified resin	50	672	609	607	963	2305	2958

Compression strength results are presented in Figure 2. Compression strength of the samples decreased as the amount of pristine resin in the samples increased whereas with gasified resin, such a behaviour was not observed.

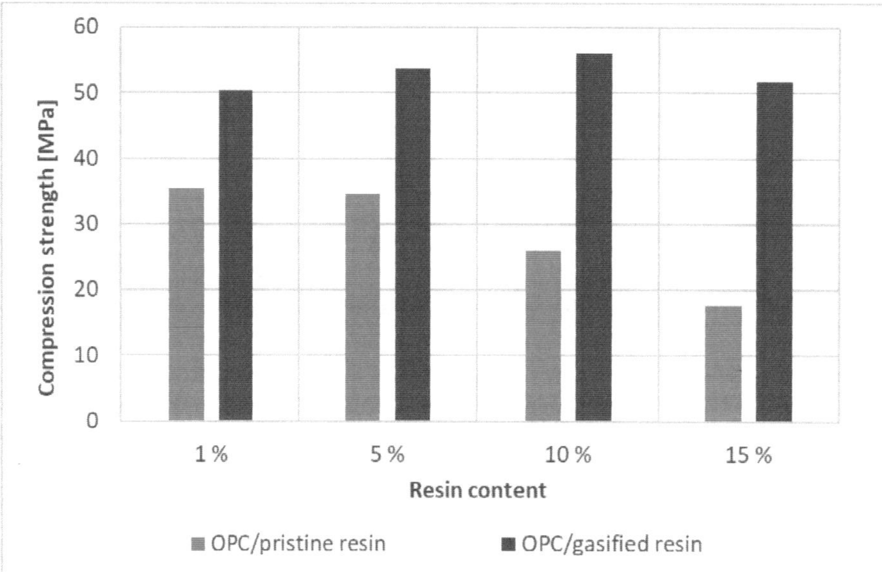

Figure 2. Compression strengths of the OPC/pristine resin, OPC/gasified resin.

4. Discussion

The United States Nuclear Regulatory Commission has defined a limit for leachability of solidified low-level radioactive wastes by a short-term test procedure. Materials with average apparent diffusion coefficients lower than 10^{-6} cm^2/s are suitable for encapsulation of low-level radioactive waste. Each studied material combination fulfilled the boundary limit requirement.

According to our results, gasification of IXR is a beneficial technique with respect to apparent diffusion coefficients and compression strengths. Average apparent diffusion coefficients decreased when the resin was gasified. If the OPC binder was - replaced with BFS or MK, the decrease was even greater. According to calculations, MK matrix and gasified resin enabled incorporation of almost 800 times more cesium in the sample, compared to traditionally used OPC/pristine resin combination. Addition of gasified resin did not decrease the compression strength of the samples within the studied range. Compression strength has been a limiting factor with pristine resin.

The maximum dosing of the gasified resin was not defined in this study. It remained unclear what would be the limiting factor for developed encapsulation system. Theoretically, limiting factors could be one of the following:

- Apparent diffusion coefficient
- Mechanical properties
- Radioactivity

Apparent diffusion coefficient will become a limiting factor at some point. According to calculations, such point will be around 5 times higher dosing in OPC matrix, 9 times in BFS and 800 times in MK.

Compression strength studies did not present decrease in compression strength up to 15% dosing of gasified resin, which is in region of the proposed dosing. It is likely that mechanical properties will not be the limiting factor for gasified resins within the studied range. Radioactivity of the encapsulated samples would increase similarly to the dosing. It is possible that increasing the activity of the sample by 800 times could change the wastes classification from LILW to HLW.

Although the ultimate limits for the studied system was not achieved, two interesting points were studied. First point is the dosing corresponding to cesium content 35% of pristine resin where the Cesium dosing corresponds to cesium content of IXR in the bulk density. According to results, MK encapsulation generated Cs leachate concentrations below those measured with OPC/pristine resin. Another interesting point was cesium content corresponding to 50% of pristine resin. Encapsulation to solid glass phase via melting has typically pristine resin content of 50%. MK matrix enabled gasified resin corresponding to 50% pristine resin content. Compression strength of the sample was good and Cs leachate concentrations were below those obtained for conventional OPC/pristine resin wasteforms.

5. Conclusions

This study demonstrated that gasification of the IXR would be a beneficial treatment to reduce the total volume of LILW requiring disposal. Encapsulation of gasified resin increased sample's compression strength and reduced the apparent diffusion coefficient of Cs tracer compared to pristine resin sample. Changing the binder from OPC to BFS or MK reduced apparent diffusion coefficient even further. Specifically, the use of MK decreased the rate of Cs diffusion from the sample and some of the leachate concentrations were even below detection limit. Calculated average apparent diffusion coefficients enabled comparison of the encapsulated Cs content assuming identical leachate concentrations. According to calculations, use of the MK matrix would enable 800 times higher initial Cs concentration compared to conventional OPC/pristine resin encapsulation matrices.

References

[1] Posiva Oy 2011 *Final disposal of spent nuclear fuel in Olkiluoto*

[2] Nieminen M, Laatikainen-Luntama J and Olin M 2020 Gasification-based thermal treatment of Low and Intermediate Level Waste *IOP Conf. Ser.: Mater. Sci. Eng* same issue

[3] Vienna J D 2010 Nuclear Waste Vitrification in the United States: Recent Developments and Future Options *Int. J. Appl. Glas. Sci.*, **1**, 3, pp. 309–321

[4] Vance E R and Perrera D S 2009 Geopolymers for nuclear waste immobilisation *Geopolymers - Structure, Processing, Properties and Industrial Applications* pp. 401–420

[5] ANS2003 *Measurement of leachability of solidified low level radioactive wastes by short-term procedure* American Nuclear Society, Illinois, USA, ANSI/ANS-16.1-2003

[6] SFS-EN 197-1 "Tavallisten sementtien koostumus, laatuvaatimukset ja vaatimustenmukaisuus" 2012

Acknowledgements
This project has received funding from the Euratom research and training programme 2014-2018 under grant agreement No 755480. This paper reflects only the authors' views, and the European Commission is not responsible for any use that may be made of it.

Synthesis and characterisation of high ceramic fraction brannerite (UTi_2O_6) glass-ceramic composites

M C Dixon Wilkins[1], E R Maddrell[2], M C Stennett[1] and N C Hyatt[1]*

[1] Department of Materials Science & Engineering, The University of Sheffield, Mappin Street, Sheffield, S1 3JD.
[2] National Nuclear Laboratory, Sellafield, Cumbria, UK

*n.c.hyatt@sheffield.ac.uk

Abstract. Brannerite, UTi_2O_6, glass-ceramic composites have been prepared, using UO_2 and TiO_2 as the ceramic phase precursors. A range of cold-press and sinter samples with varying glass:ceramic ratios have been prepared under argon at 1200 °C to investigate the effect of glass content on formation of brannerite. Ceramic brannerite formed well in all compositions, even at low (10% by weight) glass fractions, with UO_2 as a minor product. Three further brannerite glass-ceramics have been prepared by hot isostatic pressing to investigate the compatibility of this system to HIPing. The samples HIPed at 1200 °C form brannerite, with UO_2 as a minor phase with slightly higher abundance than in the cold-press and sinter samples.

1. Introduction

Brannerite (prototypically UTi_2O_6) is a naturally occurring mineral, notable for having a particularly high uranium content (> 55% by weight uranium) and being highly durable. Natural samples of brannerite have been found to retain the majority of their uranium inventory, even when they have been completely amorphised (*i.e.* they are metamict) and altered over time [1]. These properties make brannerite a possible candidate material for immobilisation of high actinide content nuclear wastes, as they are evidence that a brannerite host matrix may have a high waste loading and aqueous durability.

The properties of ceramic wasteforms (such as brannerite) make them excellent hosts for certain waste streams; however, their more limited chemical flexibility (especially compared to vitreous or encapsulating technologies) reduces the possible applications to those waste streams that are already reasonably pure. This lack of chemical flexibility can be remedied by the addition of a secondary vitreous phase, resulting in a glass-ceramic (where the ceramic phase crystallises from the glass during an annealing step) or glass-ceramic composite (where the glass and ceramic phases may either form at the same time, or both be pre-formed, and simply sintered together) depending on the processing route.

Glass-ceramic wasteform materials have many benefits when compared to single-phase ceramic wasteforms. Primarily is the wider range of wastes that the system may feasibly immobilise, where the inclusion of a glass phase with low actinide solubility does not preclude a suitable wasteform, as the actinide will partition into the ceramic phase. An often overlooked factor is the reduction in processing temperature caused by addition of the glass phase during synthesis. For example, uranium brannerite ceramics require temperatures of 1350 °C or higher for an extended duration to form [2–5], but various brannerite-structured ceramic phases within a glass-ceramic composite system have been shown to form at 1200 °C [6–9].

This is especially important when considering the compatibility of these materials with thermal processing techniques. In this study hot isostatic pressing (HIP or HIPing) has been used to consolidate these materials. The temperature of formation is of particular importance for HIP, as the most common and easily available canister material is stainless steel, with a maximum HIPing temperature of 1300 °C; so it is desirable to reduce the reaction temperature *e.g.* through the addition of a glass phase.

In this study UTi_2O_6 glass-ceramics have been produced by two separate methods to investigate the viability of HIP as a processing route. Starting from the component oxides (UO_2, TiO_2) and a glass precursor, a cold-press and sinter (CPS) method has been used to examine how this lowered temperature of formation may be dependent on the relative abundance of the vitreous phase, and preliminary HIP samples have been prepared as an initial proof-of-concept.

2. Experimental

2.1. Production of glass-ceramics

All samples were prepared using UO_2 and TiO_2 to form the ceramic phase, and a pre-calcined glass precursor to form the vitreous phase. The glass precursor was formed by calcining a homogeneous mixture (targeting a glass composition of $Na_2AlBSi_6O_{16}$) of the glass reagents (Na_2CO_3, H_3BO_3, Al_2O_3, and SiO_2) at 600 °C for 6 hours to remove volatile species. The ceramic components and the glass precursor were then wet milled together in isopropanol to homogenise. The resulting slurry was dried at 90 °C, and the powders retrieved for further processing.

Samples prepared by a CPS route were prepared by pressing into pellets (approximately 250 mg each) and then heat treated on a bed of coarse ZrO_2 (to prevent them from sticking to the crucibles) at 1200 °C for 12 hours, under an Ar atmosphere.

Samples prepared by HIP were prepared by first calcining the as-milled powders at 600 °C to remove any remaining volatiles. The powders were then pressed into stainless steel HIP canisters under approximately 1 ton using a hydraulic press (each canister contained approximately 25 g of material). The can lids were welded on, and the cans evacuated under vacuum at 600 °C to further ensure removal of all volatile materials. The evacuation tube was then crimped off and welded shut. The prepared cans were then HIPed at 100 MPa at the desired temperature and time.

Table 1 contains the as-batched compositions of all samples, as well as the U:Ti molar ratio and details of all heat treatments.

2.2. Characterisation

All samples were characterised using powder X-ray diffraction (XRD, Bruker D2 Phaser, Ni-filtered Cu Kα), and scanning electron microscopy (SEM, Hitachi TM3030 with Bruker Quantax EDX system). Samples were prepared for SEM by mounting and polishing to an optical finish using successive grades of abrasive paper and diamond suspensions.

Table 1. As-batched compositions (% by weight) of brannerite glass-ceramics, along with details of their heat treatment.

Sample ID	Glass	UO_2	TiO_2	U:Ti molar ratio	Heat treatment
CPS50	50%	31.42%	18.58%	1:2	1200 °C, 12 h, Ar
CPS60	40%	37.70%	22.30%	1:2	
CPS70	30%	43.98%	26.02%	1:2	
CPS80	20%	50.27%	29.73%	1:2	
CPS90	10%	56.55%	33.45%	1:2	
HIP1100	50%	31.42%	18.58%	1:2	HIP, 1100 °C, 24 h
HIP1200a	50%	31.42%	18.58%	1:2	HIP, 1200 °C, 4 h
HIP1200b	48.75%	30.63%	20.62%	1:2.277	

3. Results and discussion

3.1. The effect of glass-ceramic ratio on the formation of brannerite

The compositions that were processed by a CPS route show broadly the same phase assemblage. The XRD patterns all confirm that brannerite, UTi_2O_6, was formed as the predominant crystalline product, along with a small amount of unreacted UO_2 (less than 2% by weight). As the ratio of glass to ceramic is changed from 50:50 (by weight) to 10:90 the relative amount of UO_2 increases, but still remains a minor phase (see Figure 2).

The microstructures seen in the SEM micrographs are in good agreement with the XRD phase assemblages. As the glass content decreases from sample CPS50 to sample CPS90, the observed microstructures show increasing amounts of brannerite, with a corresponding increase in the amount of UO_2 also observed. In all samples, the majority of UO_2 is seen as small regions encapsulated within grains of brannerite, with a smaller amount of UO_2 particles held directly within the glass phase. Macroscale pores are visible in all samples; these are most likely due to incomplete calcination of the glass precursor causing some volatiles to only be removed during the final heat treatment at 1200 °C, rather than during the glass precursor forming pre-calcination at 600 °C (see Figure 1).

The presence of excess unreacted UO_2 is unsurprising; previous reports of brannerite glass-ceramics in the literature have noted that addition of superstoichiometric amounts of the titanium starting material is necessary to prevent formation of these regions of UO_2. This may be due to one or more different factors, depending on choice of processing route. Dissolution of TiO_2 in the glass phase during heat treatment may prevent full reaction of the uranium, loss of titanium during processing will also have the same effect (especially relevant for alkoxide/nitrate based routes, where titanium isopropoxide will readily hydrolyse in air during sample preparation), or some level of A-site non-stoichiometry in the final brannerite phase (which has previously been observed in natural and synthetic brannerites alike [1,10]).

As the relative glass content decreases from 50% to 10% by weight, the temperature dependence of brannerite formation appears to be unaffected. This is evidenced by the majority crystalline product in all samples being brannerite, with no remaining crystalline TiO_2 observed in the diffraction patterns or SEM micrographs. This shows that this glass-ceramic composite system forms well at processing temperatures achievable in standard stainless steel HIP canisters, and so three preliminary HIPed samples have also been prepared to confirm this.

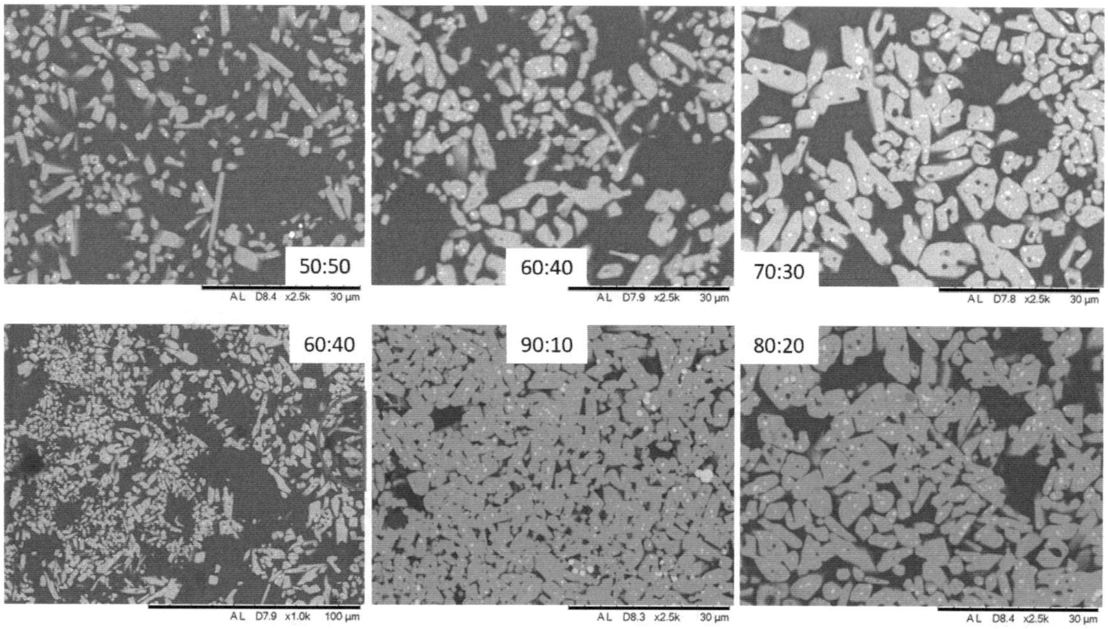

Figure 1. SEM micrographs of UTi_2O_6 glass-ceramics with different ceramic:glass ratios, as indicated. The white phase is UO_2, light grey is UTi_2O_6, and dark grey is glass. Note that the bottom left image is at a different magnification, and has a large region of porosity (as indicated in red).

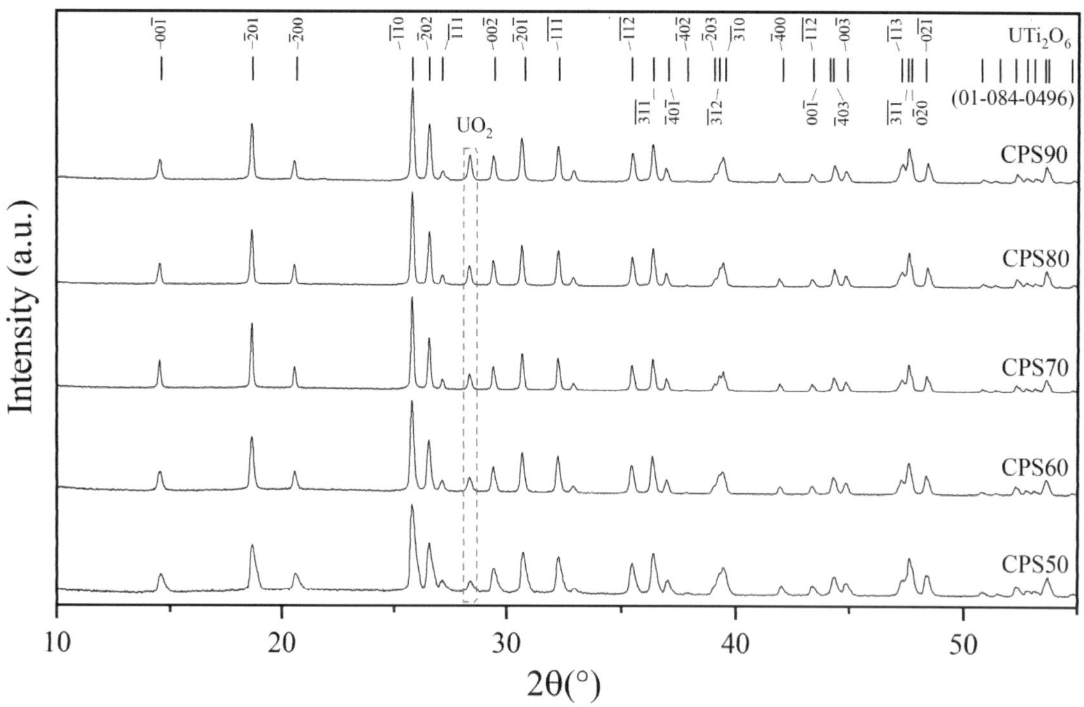

Figure 2. XRD patterns of CPS brannerite glass-ceramics with differing ceramic:glass ratio. The ceramic:glass ratio of each pattern is indicated, and the reflections of UTi_2O_6 are marked at the top.

3.2. HIPed brannerite glass-ceramic composites

Three HIP samples were prepared, one at 1100 °C and two at 1200 °C. The first sample, HIP1200a has the same composition as the sample CPS50, and was also processed at 1200 °C (however it was only heated at 1200 °C for four hours, rather than the 12 hours of the CPS samples). The main phases observed by XRD and SEM are brannerite and UO_2; however, the relative amount of UO_2 is higher (approximately 5% by weight) than that of the sample CPS50 (see Figure 3). As mentioned above, previous reports have shown that hyperstoichiometric amounts of TiO_2 can be necessary to prevent formation of free UO_2, so sample HIP1200b was batched from the same precursors as HIP1200a but with the addition of a further 2.5 wgt% TiO_2 prior to the final homogenisation step. The phases observed by XRD and SEM match those of HIP1200a; however, the addition of excess TiO_2 reduced the relative amount of UO_2 in the phase assemblage. A small peak assignable to TiO_2, rutile, is also observed, suggesting that addition of further TiO_2 would not cause complete reaction of the UO_2.

The sample HIP1100 was made to investigate if the temperature of reaction could be further reduced (with a corresponding increase of the time at temperature to encourage full reaction) whilst still producing a high brannerite fraction glass-ceramic composite. The phase assemblage as observed by XRD and SEM consists of brannerite and UO_2, the same as the samples processed at 1200 °C; however, the relative amount of UO_2 is evidently higher than for all other samples.

It is unclear why more UO_2 is observed in the HIPed samples when compared to CPS samples of the same composition. A possible explanation is offered by a report that the brannerite structure is unstable in reducing conditions

[5], such as those exerted by the stainless steel HIP canister. This is supported by the observation of higher amounts of unreacted UO_2 at the can-sample interface in the HIPed samples. However, samples of pure ceramic UTi_2O_6 have been formed under reducing atmospheres (24 hours at 1350 °C in 5% H_2/95% N_2 atmosphere, unpublished work), so this may not offer a complete explanation of the phenomenon.

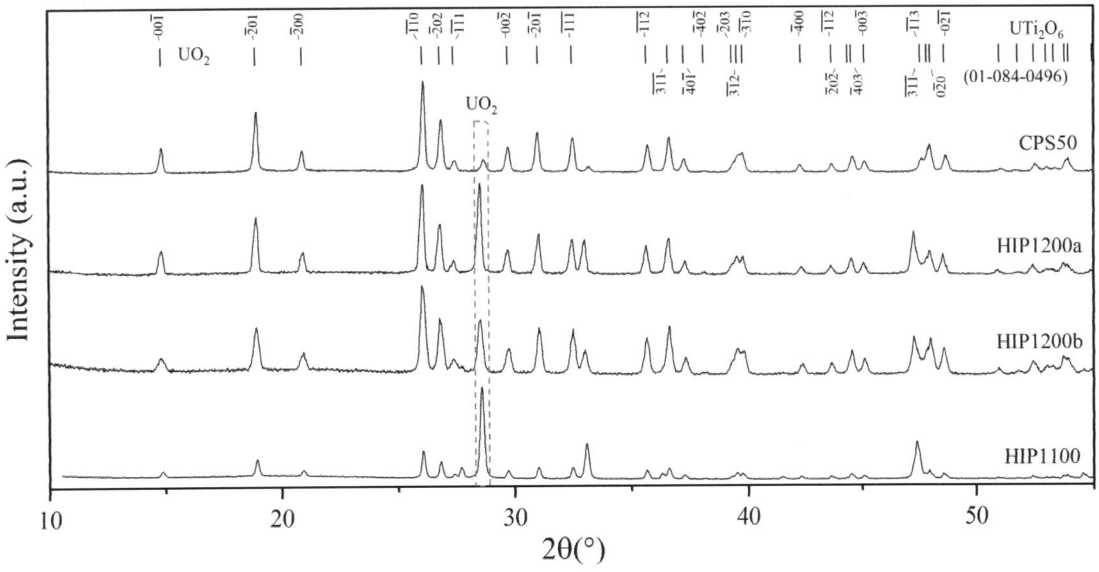

Figure 3. XRD patterns of brannerite glass-ceramics HIPed under different conditions. The conditions of each HIP sample are indicated and the reflections of UTi_2O_6 are marked at the top.

4. Conclusion

A range of UTi_2O_6 glass-ceramic composites have been synthesised, with a focus on investigating the effect of the glass:ceramic ratio on the temperature of formation of brannerite. When processed via a

CPS route at 1200 °C, UTi_2O_6 formed as the major crystalline phase for glass:ceramic ratios from 50:50 to 10:90, with UO_2 as a secondary product. When HIPed at 1200 °C, brannerite is still the majority product; however, the relative amount of UO_2 increases when compared to a CPS sample of the same as-batched composition. The mechanism causing this is not clear. When the temperature of reaction is further reduced to 1100 °C, significantly less brannerite is formed, with a much higher proportion of UO_2 observed in the final product.

As expected from the synthesis of brannerite glass-ceramic composites in the literature, UTi_2O_6 forms well in glass at 1200 °C, below the temperatures required for synthesis of a full ceramic from component oxides. Further work required to fully understand the extent and exact cause (although it is most likely caused by the glass phase lowering the barrier to diffusion of the ceramic phase reactants) of this effect is ongoing, with a focus on in-situ high temperature diffraction techniques.

5. References

[1] Lumpkin G R, Leung S H F and Ferenczy J 2012 *Chem. Geol.* **291** 55.
[2] Bailey D J, Stennett M C and Hyatt N C 2016 *Procedia Chem.* **21** 371.
[3] Bailey D J, Stennett M C and Hyatt N C 2017 *MRS Adv.* **2** 557.
[4] James M, Carter M L and Watson J N 2003 *J. Solid State Chem.* **174** 329.
[5] Vance E R, Watson J N, Carter M L, Day R A and Begg B D 2001 *J. Am. Ceram. Soc.* **84** 141.
[6] Zhang Y, Kong L, Karatchevtseva I, Aughterson R D, Gregg D J and Triani G 2017 *J. Am. Ceram. Soc.* **100** 4341.
[7] Zhang Y, Karatchevtseva I, Kong L, Wei T and Zhang Z 2018 *J. Am. Ceram. Soc.* **101** 5219.
[8] Zhang Y, Wei T, Zhang Z, Kong L, Dayal P and Gregg D J 2019 *J. Am. Ceram. Soc.* **102** 7699.
[9] Dixon Wilkins M C, Stennett M C and Hyatt N C 2019 *MRS Adv.* in press.
[10] Stennett M C, Freeman C L, Gandy A S and Hyatt N C 2012 *J. Solid State Chem.* **192** 172.

Acknowledgment

This research was sponsored in part by EPSRC under grant references EP/N017870/1, EP/S011935/1 and EP/S01019X/1. MCDW is grateful to EPSRC and NDA for provision of a doctoral training award. This research utilised the HADES / MIDAS Facility, established with financial support from the UK Department for Business, Energy and Industrial Strategy and EPSRC under grant reference EP/T011424/1.

IOP Conf. Series: Materials Science and Engineering **818** (2020) 012019 doi:10.1088/1757-899X/818/1/012019

Ce and U speciation in wasteforms for thermal treatment of plutonium bearing wastes, probed by L_3 edge XANES

D J Bailey,[1] M C Stennett,[1] S K Sun,[1] L J Gardner,[1] S A Walling,[1] C L Corkhill,[1] B Ravel,[2] and N C Hyatt.[1]*

[1]Department of Materials Science & Engineering, The University of Sheffield, Mappin Street, Sheffield, S1 3JD.
[2]National Institute of Standards and Technology, Gaithersburg, Maryland 20899, USA.

*n.c.hyatt@sheffield.ac.uk

Abstract. X-ray absorption spectroscopy was applied to understand the speciation of elements relevant to the immobilisation and disposal of radioactive plutonium bearing wastes, utilizing Ce as a Pu surrogate. Ce L_3 XANES (X-ray Absorption Near Edge Structure) characterisation of a crystallised glass material produced by cold crucible plasma vitrification, at demonstration scale, evidenced incorporation as Ce^{3+} within the glass phase, providing an important validation of laboratory scale studies. U and Ce L_3 XANES investigation of brannerite ceramics, $U_{0.9}Ce_{0.1}Ti_2O_6$, synthesized under oxidizing, neutral and reducing conditions, established the charge compensation mechanism as incorporation of Ce^{3+} through formation of U^{5+} and/or U^{6+} In each of these examples, X-ray Absorption Spectroscopy has provided a pivotal understanding of element speciation in relation to the mechanism of incorporation within the host wasteform intended for geological disposal.

1. Introduction

The UK is engaged in a programme of decommissioning, clean up and disposal of legacy facilities and materials from over five decades of nuclear fuel cycle activities. This programme will extend over more than 120 years and cost in excess of £115 billion [1]. The UK radioactive waste inventory comprises a spectrum of plutonium bearing wastes, which include plutonium residues and plutonium contaminated materials (PCM), generated by reprocessing of nuclear fuel [2-8]. The inventory of plutonium and Mixed Oxide (MOX) fuel residues is of the order of 0.5 t and includes material from which plutonium recovery is considered uneconomic [2], including PuO_2 which is heavily contaminated by chlorine as a result of long term storage in PVC packaging which has degraded. The inventory of untreated PCM waste comprises in excess of 20,000 m^3 on the Sellafield site alone [5].

A tool box of thermal treatment technologies is under development for conditioning of plutonium / MOX residues and PCM wastes, as well as other complex higher activity wastes, which are incompatible with cement encapsulation [2-13]. Primary drivers for the application of thermal treatment processes include the reduced volume, improved passive safety, and superior long term stability of the waste form products [3]. These benefits arise from destruction of the organic waste fraction, oxidation of the metal waste component, and evaporation of entrained water, in tandem with the immobilisation of radioactive and chemotoxic elements within a durable ceramic, glass, or glass-ceramic material.

Here, we demonstrate the crucial role of X-ray absorption spectroscopy in understanding plutonium surrogate and contaminant speciation in wasteforms designed for immobilisation of PCM wastes and plutonium / MOX residues by thermal treatment: a crystallised glass (or slag) waste form, produced by cold crucible plasma vitrification at the demonstration scale, to immobilise PCM wastes; and a ceramic waste form, produced by cold pressing and sintering at the laboratory scale, tailored to immobilise MOX fuel residues. In each case study, the critical role of speciation, determined by X-ray Absorption Spectroscopy, in controlling the incorporation mechanisms, is emphasised.

2. Experimental

2.1 Plasma vitrified wasteforms

A simulant PCM waste, using CeO_2 as a PuO_2 surrogate, was vitrified using a twin torch plasma system, at waste loadings of 31, 40 and 54 wt% PCM waste, with glass forming additives (CaO 29.2 %, Al_2O_3 27.7 % and SiO_2 43.1 % by weight). The slag melt was maintained at ca. 1500°C in the cold crucible and quenched to form a glassy solid, as previously described [9]. Ce L_3 edge (5723.4 eV) XANES data from plasma vitrified products, and reference compounds, were acquired from polished solid specimens on Beamline 9A of the Photon Factory, Japan. The configuration utilised: Si (111) double crystal monochromator; harmonic rejection mirror; ion chambers (I_0), and 19 element Ge solid-state fluorescence detector [14]. Samples were prepared by diluting the material to be investigated in polyethylene glycol to yield a thickness of $\mu x = 1$. All XANES data were dead time corrected and processed in Athena using standard background subtraction and normalisation procedures [15].

2.2 Brannerite ceramic wasteforms

Reagents (UO_2, CeO_2, TiO_2) were combined in stoichiometric ratio with isopropanol to form a slurry and ball milled for 5 min at 30 Hz using a Fritsch Pulverisette 23. The dried slurry was calcined at 1300°C for 12 h and sintered at 1320°C for 12 h; heat treatments were performed under air, Ar or 5 % H_2 / N_2 atmosphere. U and Ce L_3 edge XANES data from brannerite ceramics and reference compounds were acquired in transmission and fluorescence mode, respectively, on the now decommissioned beamline X23A2 of the National Synchrotron Light Source, USA [16]. The configuration utilised: Si (311) double crystal monochromator; Rh harmonic rejection mirror; ion chambers (I_0, I_t, I_{ref}). U L_3 edge XANES (17166.3 eV) were acquired in transmission mode whereas Ce L_3 edge XANES (5723.4 eV) were acquired using a four element Vortex Si-drift detector in fluorescence mode with the aid of a harmonic rejection mirror. Additional U L_3 edge XANES data were acquired, using the same procedure, on the 6-BMM, National Synchrotron Light Source II, using a double crystal Si (111) monochromator. Reference compounds were synthetic $CePO_4$, CeO_2, UTi_2O_6, $U_{0.5}Y_{0.5}Ti_2O_6$, and $CaUO_4$. Samples were prepared for XANES analysis by diluting the material to be investigated in polyethylene glycol to yield a thickness of $\mu x = 1$. All XANES data were dead time corrected and processed in Athena using standard background subtraction and normalisation procedures [15].

2.3 General characterization methods

Scanning Electron Microscopy (SEM) coupled with Energy Dispersive X-ray (EDX) analysis was performed on polished and carbon coated specimens, using a Hitachi TM3030 SEM with a 15 kV accelerating voltage and a working distance of 8 mm. Compositional analysis was conducted with a Bruker Quantax EDX system. Samples were ground and sieved for powder X-ray diffraction (PXRD) using a Bruker D2 PHASER diffractometer with Ni filtered Cu Kα radiation (1.5418 Å); data were acquired between $10° \leq 2\theta \leq 70°$, with a step size of 0.02°, using a Lynxeye position sensitive detector.

3. Results and discussion

3.1 Plasma vitrified wasteforms

Cold crucible plasma arc vitrification is a candidate technology for thermal treatment of plutonium contaminated material (PCM) wastes, to produce a partially crystallised glass (or slag) waste form [9]. In this study, the speciation of Ce, as a Pu analogue, was investigated in the slag products of cold crucible

IOP Conf. Series: Materials Science and Engineering **818** (2020) 012019 doi:10.1088/1757-899X/818/1/012019

plasma arc vitrification, produced at demonstration scale, with 31, 40 and 54 wt% simulant PCM waste, and glass forming additives (CaO 29.2 %, Al_2O_3 27.7 % and SiO_2 43.1 % by weight). Details of the waste simulant composition and a comprehensive study of the slag microstructure, phase assemblage and composition were published previously [9].

Figure 1 shows, as an example, the microstructure of the 54 wt% PCM slag, which comprises two amorphous components with light and mid grey contrast; EDX showed the phase of lighter contrast (A) was depleted in Al_2O_3 and enriched in Fe_2O_3, compared to the phase of darker contrast (B). Additional crystalline phases of mullite, anorthite, alumina, and spinel were identified by PXRD, evident in the microstructure as angular dark grey features (e.g. C). EDX demonstrated Ce to partition exclusively to the amorphous phases, with no preference for either amorphous component within the margin of precision [9]. The composition and microstructure of all slag materials were similar, though with increasing amorphous fraction with decreasing waste loading, as assessed from SEM/EDX and PXRD.

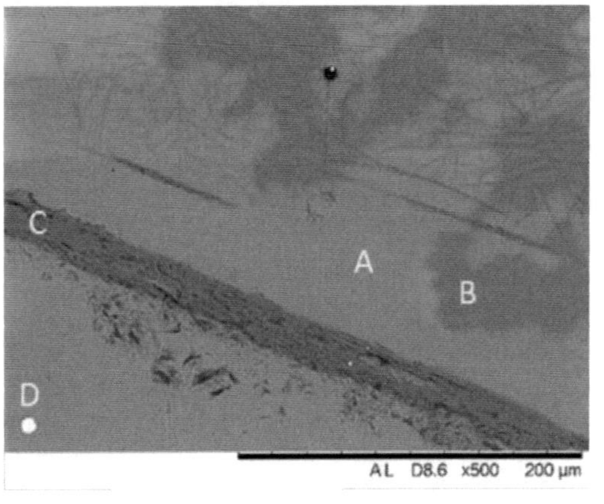

Figure 1. Microstructure (back scattered electron image) of plasma vitrified simulant PCM waste (54% waste loading by weight), comprising two amorphous components (A and B); crystalline mullite, anorthite, alumina, and spinel phases (C), and a ferrous alloy (D).

Ce L_3 edge XANES data, Figure 2, were acquired to determine the Ce speciation in the amorphous phase. The low Ce concentration (ca. 0.6(2) mol % CeO_2 determined by EDX) required data acquisition in fluorescence mode, which was limited to the XANES region due to edge overlap by trace Cr_2O_3 in the material. The Ce L_3 XANES features and edge shift of all slag compositions closely resemble those of Ce^{3+} in $CePO_4$ monazite, characterised by a single intense feature attributed to the transition from an initial $2p^64f^15d^0$ state to a $2p^54f^15d^1$ final state [17-19]. In contrast, the Ce L_3 edge XANES of Ce^{4+} in CeO_2 is characterised by a white line comprising two features of lower relative intensity due to transition from an initial $2p^64f^05d^0$ state to: (A) the $2p^54f^05d^1$ final state; and (B) the $2p^54f^15d^1L^1$ final state (where L^n denotes a ligand hole); feature C is associated with transition to a $2p^54f^25d^1L^2$ final state, whereas D is assigned to quadrupole transition [17-19]. The XANES spectra shown in Figure 5 permit straightforward fingerprinting of the Ce oxidation state. From comparison of the spectra it is apparent that the XANES data of slag fraction closely resemble that of $CePO_4$, demonstrating that Ce^{3+} is the dominant species, no significant Ce^{4+} contribution was determined by linear combination fitting of the XANES data of $CePO_4$ and CeO_2 reference compounds.

Previous Ce L_3 edge XANES characterisation of slag wasteforms produced by laboratory scale crucible melts utilising a wider spectrum of simulant PCM wastes, and a closely related glass composition, also demonstrated Ce partitioning to the glass phase exclusively as reduced Ce^{3+} [5]. The solubility of Ce_2O_3 and Pu_2O_3 in (boro)silicate glasses has been shown to be broadly similar, with an upper limit of ca. 2 mol% at 1450 °C [20]. Thus, our speciation data provide confidence of Pu incorporation within the glass component of the slag phase in the proposed plasma treatment process, since the incorporation rate is expected to be well below the anticipated solubility limit, given the evidently reducing nature of the melt.

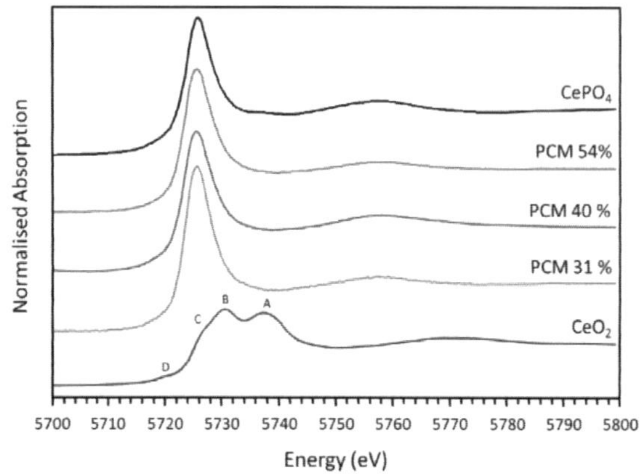

Figure 2. Ce L_3 edge XANES data acquired from plasma vitrified simulant PCM waste, with 31 wt%, 40 wt% and 54 wt% waste loading by weight, compared with CeO_2 and $CePO_4$ reference compounds, see text for detail.

3.2 Brannerite ceramic wasteforms

Brannerite, prototypically UTi_2O_6, is a potential ceramic host for residues arising from MOX (Mixed Oxide, $(Pu,U)O_2$) fuel fabrication [8, 21]. The crystal structure of brannerite comprises layers of anatase-like edge sharing TiO_6 octahedra which are linked by UO_6 octahedra that also share edges [8, 21]. In this exploratory study, we investigated the synthesis of brannerites formulated $U_{0.9}Ce_{0.1}Ti_2O_6$, under air, argon and 5 % H_2 / N_2 atmosphere, with Ce as a Pu surrogate at realistic concentration. Representative microstructures of ceramics sintered at 1320 °C for 12 h are shown in Figure 3: synthesis under air (Fig. 3a) afforded U_3O_8 as the major phase (A) with minor brannerite (B) and rutile (R); whereas, synthesis under argon and 5 % H_2 / N_2 (Fig 3b and 3c, respectively) produced brannerite as the major phase (B), with minor rutile (R) and trace UO_2 (C). Ce was determined to partition exclusively to the brannerite phase by EDX spot analysis and mapping, Figure 3. The phase assemblages were confirmed by PXRD.

Ti K, Ce L_3 and U L_3 XANES were exploited to understand the charge compensation mechanisms and phase assemblages in $U_{0.9}Ce_{0.1}Ti_2O_6$ brannerite ceramics. Ce L_3 edge XANES of the brannerite ceramics, Figure 4, revealed reduction of Ce^{4+} to Ce^{3+} under all processing conditions, by comparison with data from $CePO_4$ and CeO_2 (reference compounds for Ce^{3+} and Ce^{4+}, respectively). The U L_3 XANES of brannerites synthesised in 5 % H_2 / N_2 and argon showed an edge shift comparable to that of the UTi_2O_6 reference, Figure 5, whereas that of the brannerite sintered under air showed an edge shift between $Y_{0.5}U_{0.5}Ti_2O_6$ and $CaUO_4$. Linear regression of oxidation state versus edge shift for the reference compounds enabled the average U oxidation state of the brannerite compounds to be determined as 5.4+ for the air processed material and 4.1+ and 4.0+ for the samples produced under argon and 5 % H_2 / N_2, respectively (± 0.1). Ti K edge XANES data (not shown) revealed the Ti oxidation state in the brannerites to be 4.0+, within precision, irrespective of the processing atmosphere.

Near single phase $U_{0.9}Ce_{0.1}Ti_2O_6$ brannerite could evidently only be synthesised under argon or 5 % H_2 / N_2 atmosphere. The observed reduction of Ce^{4+} to Ce^{3+} implies charge compensation by marginal oxidation of U^{4+}, which was within the uncertainty of the U oxidation state determined by U L_3 XANES. Under air atmosphere the average U oxidation state was 5.4+, consistent with the coexistence of U_3O_8, and the Ce rich brannerite stoichiometry determined by EDX thus implies charge compensation by U^{5+} and/or U^{6+}.

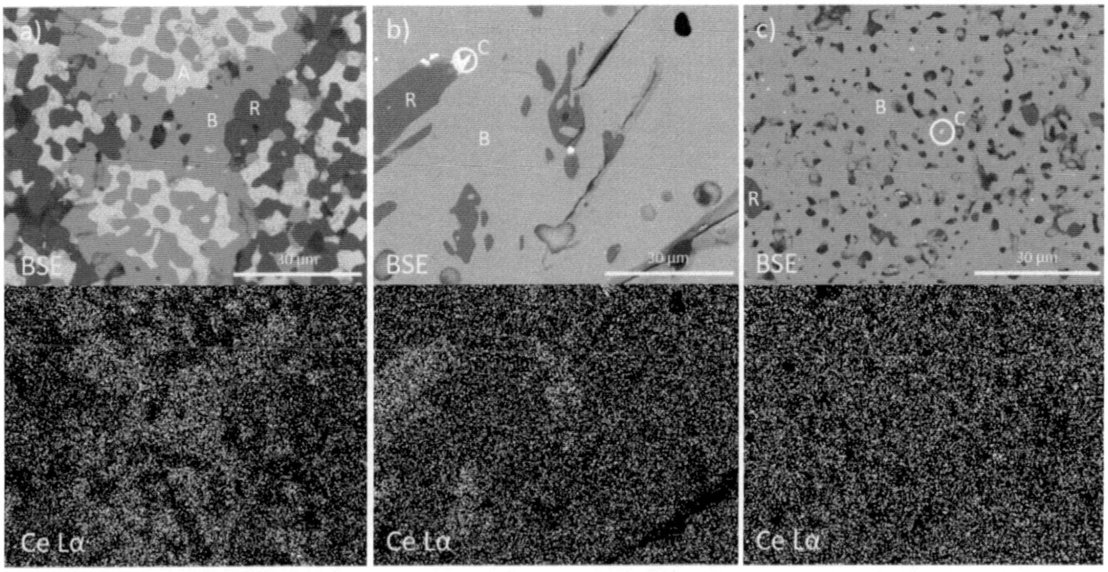

Figure 3. Microstructures (back scattered electron images) of brannerite ceramics with formula $U_{0.9}Ce_{0.1}Ti_2O_6$ synthesised under a) air, b) argon and c) 5 % H_2 / N_2 atmosphere; A = U_3O_8, B = $(U,Ce)Ti_2O_6$, R = TiO_2, Rutile; C = UO_2. Note overlap of Ti Ka and Ce La emission lines gives rise to apparent Ce incorporation in rutile phase.

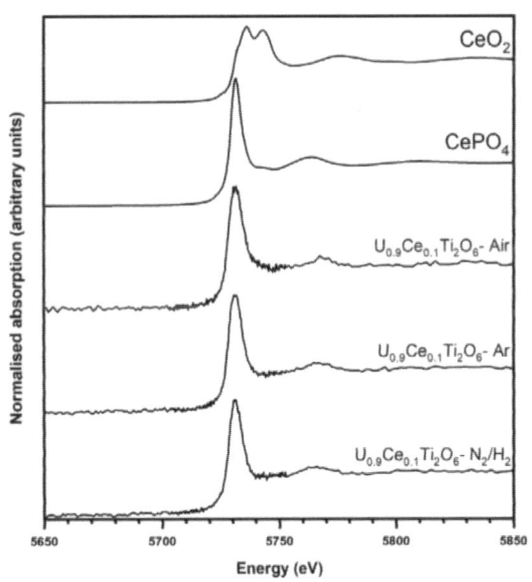

Figure 4. Ce L_3 XANES of brannerite ceramics sintered under air, argon, or 5% H_2 / N_2 atmosphere, plus reference compounds $CePO_4$ and CeO_2.

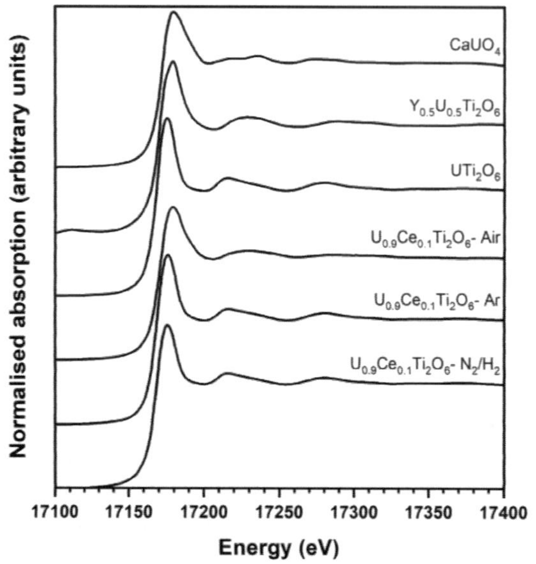

Figure 5. U L_3 XANES of brannerite ceramics sintered under air, argon, or 5 % H_2 / N_2 atmosphere, plus reference compounds UTi_2O_6, $U_{0.5}Y_{0.5}Ti_2O_6$, and $CaUO_4$.

4. Conclusions

The Ce L_3 edge XANES data of slag wasteforms produced by cold crucible plasma vitrification, provide a useful and essential verification of Ce partitioning and reduction in a demonstration scale cold crucible processing, which was not sensitive to waste loading. The common Ce^{3+} speciation achieved in demonstration and laboratory scale slag wasteforms produced from thermal treatment of simulant PCM waste, results in exclusive partitioning to the glass phase under the reducing melt conditions. This observation provides important confidence that laboratory scale crucible melts provide a useful guide to surrogate partitioning behaviour in demonstration and full scale processing, in early stage research and development.

We have shown that near single phase brannerite ceramics may be fabricated by a cold-press and sinter process under neutral to reducing atmosphere, as effective candidate waste forms for MOX residue immobilisation, with waste loading of up to ca. 63 wt%. The ability of the brannerite structure to self charge compensate for potential reduction of Pu^{4+} to Pu^{3+} (as demonstrated by the Ce analogue), through oxidation of U^{4+} to U^{5+} and/or U^{6+} is an attractive feature and would mitigate against variations in U / Pu feed composition within solid solution limits (as could be expected with residues of variable Pu content). Overall, this is a promising approach since the cold-press and sinter processing technology is already implemented for MOX fuel manufacture and a fuel processing plant could conceptually be repurposed for an immobilisation mission at the end of fuel production operations.

5. References

[1] Nuclear Decommissioning Authority, 2016. Strategy, Effective from April 2016.

[2] Scales, C R, Maddrell, E, Gawthorpe, N, Begg, B D, Moricca, S, Day, R A and Stewart M W 2006 *Waste management '06, WM'06*, Feb 26- Mar. 2, 2006, at Tucson AZ, USA, 6232.

[3] Thornber, S M, Heath P G, Maddrell E R, Stennett M C and Hyatt N C 2016 *MRS Advances* **1** 4269.

[4] Thornber S M, Heath P G, Da Costa G P, Stennett M C and Hyatt N C 2017 *J. Nucl. Mater.* **485** 253.

[5] Hyatt N C, Schwarz R R, Bingham P A, Stennett M C, Corkhill C L, Heath P G, Hand R J, James M, Pearson A and Morgan S 2017 *J. Nucl. Mater.* **444** 186.

[6] Thornber S M, Stennett M C, Vance E R, Chavara D T, Watson I, Jovanovic M, Davis J, Gregg D and Hyatt N C 2018 *MRS Advances* **3** 1065.

[7] Hyatt N C 2017 *Energy Policy* **101** 303.
[8] Bailey D J, Stennett M C, Ravel B, Grolimund D and Hyatt N C 2018 *RSC Advances* **8** 2092.
[9] Hyatt N C, Morgan S, Stennett M C, Scales C R and Deegan D 2007 *MRS Symp. P.* **985** 393.
[10] Bingham P A, Hand R J and Hyatt N C 2012 *Glass Technol. Part A* **53** 83.
[11] Bingham P A, Hyatt N C, Hand R J and Forder S D 2013 *Glass Technol. Part A* **54** 1.
[12] Hyatt N C and James M 2013 *Nucl. Eng. Inter.* **58** 10.
[13] Heath P G, Stewart M W A, Moricca S and Hyatt N C 2018 *J. Nucl. Mater.* **499** 233.
[14] Nomura M and Koyama A 1999 *J. Synchrotron Rad.* **6** 182.
[15] Ravel B and Newville M 2005 *J. Synchrotron Rad.* **12** 537.
[16] Long G G, Allen A J, Black D R, Burdette H E, Fischer D A, Spal R D and Woicik, J C 2001 *J. Res. Nat. Inst. Stand. Tech.* **106** 1141.
[17] Bianconi A, Marcelli A, Dexpert H, Karnatak R, Kotani A, Jo T and Petiau J 1987 *Phys. Rev. B* **35** 806.
[18] Soldatov A V, Ivanchenko T S, Della Longa S, Kotani A, Iwamoto Y and Bianconi A 1994 *Phys. Rev. B* **50** 5074.
[19] Stennett M C, Freeman C L, Gandy A S and Hyatt, N C 2012 *J. Solid State Chem.* **192** 172.
[20] Lopez C, Deschanels X, Bart J M, Boubals J M, Den Auwer C and Simoni E 2003 *J. Nucl. Mater.* **312** 76.
[21] Bailey D J, Stennnett M C and Hyatt N C 2016 *Procedia Chem.* **21** 371.

Acknowledgment
This research was sponsored in part by EPSRC under grant references EP/L015390/1, EP/R512175/1, EP/N017374/1, EP/P013600/1, and EP/S01019X/1. This project received funding from the Euratom Research and Training Programme 2014-18, (Grant Agreement No. 755480 and the 7th Framework programme (Grant agreement No. 323300). This research accessed beamline BMM of the National Synchrotron Light Source II, a U.S. Department of Energy (DOE) Office of Science User Facility Operated for the DOE office of Science by Brookhaven National Laboratory under Contract No. DE-SC0012704. This research utilised the HADES / MIDAS Facility, established with financial support from the UK Department for Business, Energy and Industrial Strategy and EPSRC under grant reference EP/T011424/1.

IOP Conf. Series: Materials Science and Engineering **818** (2020) 012020 doi:10.1088/1757-899X/818/1/012020

Laboratory Based X-ray Absorption Spectroscopy of Iron Phosphate Glasses for Radioactive Waste Immobilisation: A Preliminary Investigation

L M Mottram,[1] M C Stennett,[1] S K Sun[1] and N C Hyatt[1]*

Department of Materials Science & Engineering, The University of Sheffield, Mappin Street, Sheffield, S1 3JD.

*n.c.hyatt@sheffield.ac.uk

Abstract. We report the application of laboratory based X-ray absorption spectroscopy to the speciation of Fe in iron phosphate glasses prepared by conventional and microwave melting. Analysis of the weak pre-edge features in Fe K-edge XANES data demonstrated glasses produced by microwave melting to have a higher fraction of reduced Fe^{2+} species, since microwave melts do not have sufficient time to equilibrate with the prevailing oxygen partial pressure, compared to counterparts produced by conventional melting. Furthermore, our laboratory XANES data are consistent with the formation of octahedral Fe^{2+} at the expense of tetrahedral Fe^{3+} species, with increasing Fe^{2+} content. These findings are consistent with the previous findings of our ^{57}Fe Mossbauer study, synchrotron XANES data, and current understanding of the structure of iron phosphate glasses, and demonstrate the utility of laboratory based XANES for routine speciation of Fe in these and other materials.

1. Introduction

The baseline treatment option for intermediate level wastes in the UK is encapsulation in an ordinary portland cement matrix, combined with super-compaction, where appropriate. However, there is a growing appreciation that some wastes, such as reactive metals and organic materials, are not compatible with this approach [1-3]. Furthermore, cement encapsulation technology results in a dilution of the radioactive waste feed and a projected increase in the packaged waste volume by typically 30%. In principal, thermal treatment technologies challenge these shortcomings, through at least partial oxidation of metallic and organic waste components, to produce a passively safe product, with retention of radioactive and chemotoxic elements [3-10].

Iron phosphate glasses are one matrix of interest for the vitrification of some challenging radioactive wastes [11-13]. The $40Fe_2O_3$–$60P_2O_5$ (mol%) glass composition, has been extensively studied for radioactive waste immobilisation, due to the high incorporation rate of elements which have low solubility in borosilicate melts (e.g. F, Cl, S and Bi), good chemical durability, and reasonable processing temperature and melt viscosity [14-16].

X-ray absorption and ^{57}Fe Mossbauer spectroscopy studies of iron phosphate glasses have demonstrated the Fe^{2+} / Σ Fe ratio to be sensitive to melt composition and processing conditions (temperature, time and atmosphere) [16-19]. This redox ratio is determined by the equilibrium: $2\,FeO_{(melt)} + \frac{1}{2}O_{2\,(g)} = Fe_2O_{3\,(melt)}$, and, in the case of the $40Fe_2O_3$–$60P_2O_5$ glass composition, up to 60% of the total Fe may be present as Fe^{2+} [17]. An increased fraction of Fe^{2+} is reported to be both detrimental to glass forming ability and radiation stability [16, 19-22], whereas chemical durability is reported to be insensitive to the fraction of Fe^{2+} [18]. The Fe^{2+} / Σ Fe ratio also governs the electrical conductivity of Fe_2O_3-P_2O_5 glass melts, which is an important consideration for some melter designs, in the context of application to radioactive waste treatment [19].

From the preceding discussion, it will be appreciated that routine determination of Fe^{2+} / Σ Fe ratio in Fe_2O_3-P_2O_5 glasses is of importance for formulation and development of such waste forms. ^{57}Fe Mossbauer spectroscopy has been widely exploited for the determination of Fe^{2+} / Σ Fe ratio, according to the distinctive isomer shift and quadrupole splitting of Fe^{2+} and Fe^{3+} species, which are also sensitive to co-ordination number [13, 16-18, 23]. However, the determination of average Fe co-ordination number is complicated by the overlap of ^{57}Fe isomer shift and quadrupole splitting ranges, particularly for amorphous materials [13, 23]. Whilst ^{57}Fe Mossbauer spectroscopy can conveniently be applied in a laboratory environment, data acquisition is dependent on ^{57}Fe concentration, ^{57}Co source activity, and sample characteristics, which translate to data acquisition time of days per sample.

Fe K-edge X-ray absorption spectroscopy has also been applied to determination of Fe^{2+} / Σ Fe ratio in Fe_2O_3-P_2O_5 glasses. The weak pre-edge features in the Fe K-edge X-ray Absorption Near Edge Structure (XANES), located ca. 15–20 eV below the edge step, have been shown to be a sensitive probe of Fe speciation [24-29]. These features are associated with 1s – 3d (quadrupole) and/or 1s – 4p (dipole) electronic transitions in the Fe absorber atom and, therefore, are sensitive to both the electronic configuration and symmetry, from which oxidation state and co-ordination number are inferred. Wilke *et al.* developed a systematic methodology to extract the normalised pre-edge features from Fe K-edge XANES data and determination of the centroid energy position and total integrated intensity, which were effectively correlated with Fe oxidation state and co-ordination number [27, 28]. Conventionally, acquisition of XANES data has required access to a synchrotron radiation beamline, which is highly competitive, time limited, and not immediate. Recently, however, we demonstrated the first proof of concept Fe K-edge XANES speciation studies, using a commercially available laboratory X-ray spectrometer, from analysis of weak pre-edge features [30]. This approach enables routine Fe speciation in materials with data acquisition times of only a few hours. Here, we build on this recent study to demonstrate the potential for Fe redox determination in $40Fe_2O_3$–$60P_2O_5$ glasses prepared by conventional and microwave melting. Our preliminary results demonstrate that laboratory based Fe K-edge XANES has sufficient resolution and sensitivity to routinely determine Fe^{2+} / Σ Fe ratio in iron phosphate glasses.

2. Experimental

2.1. Synthesis of iron phosphate glasses

Iron phosphate glasses of $40Fe_2O_3$–$60P_2O_5$ (mol%) composition were prepared by conventional and microwave heating, as previously described [23, 31, 32]. Conventional melting used a stoichiometric mixture of Fe_3O_4 and P_2O_5 (PO) or $NH_4H_2PO_4$ (AP), in covered silica crucibles, heated at 1150°C for 5h in air. Microwave melting used a stoichiometric mixture of Fe_3O_4 and P_2O_5 or $NH_4H_2PO_4$, in covered silica crucibles, heated for 12 minutes at 800W power in a domestic microwave oven (DMO). Fe_3O_4 was selected as the iron source since it is known to strongly couple to the 2.45 GHz frequency of domestic microwaves, resulting in rapid self heating. After approximately 20 seconds of microwave irradiation a dull red glow emanating from the crucible was observed through the silica lid. The intensity of the glow increased until, after approximately 1 minute, an incandescent red glow could be clearly observed. This continued until the DMO was switched off. The peak temperature of melting was estimated to be 1150°C, using a grounded thermocouple. Glasses were quenched in air and confirmed to be amorphous by X-ray diffraction.

2.2. Reference compounds

Fe K-edge XANES data were acquired from four selected reference compounds. $NaFeSi_2O_6$ (aegirine) and Fe_2SiO_4 (fayalite) are characterised by Fe^{3+} and Fe^{2+}, respectively, in octahedral FeO_6 co-ordination (CN = 6); whereas $FePO_4$ and $Fe_{1.5}Mg_{0.5}Al_{9.1}Si_{3.9}(OH)_2$ (staurolite) are characterised by Fe^{3+} and Fe^{2+}, respectively, in tetrahedral FeO_4 co-ordination (CN = 4) [33-36]. Polycrystalline $FePO_4$ and Fe_2SiO_4

were, respectively, synthesised by: solid state reaction of Fe_2O_3 and $NH_4H_2PO_4$ at 900 °C in air for 16 h; and solid state reaction of stoichiometric quantities of Fe, Fe_2O_3 and SiO_2 at 750 °C for 48 h in a sealed, evacuated, quartz tube. Mineral specimens of aegirine (Langesund Fjord, Norway), and staurolite (Georgia, USA), were provided from our own collection. All materials were confirmed to be single phase by powder X-ray diffraction. The expected Fe speciation was verified by ^{57}Fe Mossbauer analysis.

2.3. Fe K-edge XANES

Fe K-edge XANES data were acquired on an EasyXAFS XES100 spectrometer, Figure 1, based on the design of Seidler *et al.*, equipped with an air cooled X-ray tube operating at 25 kV and 4 mA, with a Hitatchi Vortex Silicon Drift Detector (SDD) [37-40]. The spectrometer was configured with a Ge (620) SBCA to scan the energy range 6970 – 7340 eV, with a step size of 0.25 eV and constant count time of 4s / point in the XANES region (total scan time 30 min per data set). The energy resolution of the SDD is ca. 140 eV, enabling rejection of the harmonic content of the incident beam and background scatter. A He flight-path was used to minimise air scatter and absorption. Transmission data were acquired with ($I_t(E)$) and without the sample ($I_0(E)$), using the same scan parameters. The Ge (620) monochromator was aligned using the clock angle procedure of Mortensen and Seidler, to account for the crystal miscut [41]. Data integration time was typically 4h per spectrum.

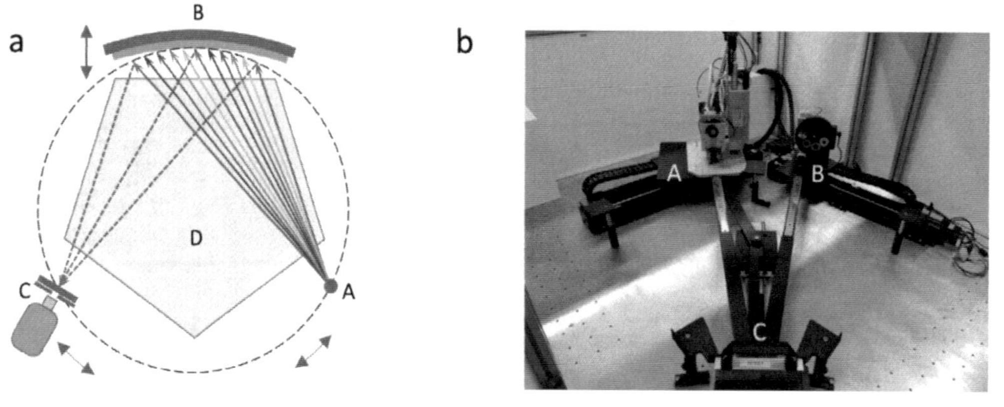

Figure 1. a) Schematic representation of Rowland circle geometry for laboratory transmission X-ray Absorption Spectroscopy, as described in the text, showing: broadband low power X-ray source (A); spherically bent crystal analyser (B); sample / detector assembly (C); and helium filled beam path with kapton windows (D); **b)** Photograph of EasyXAFS XES100 spectrometer arrangement shown in Fig. 1a, with helium filled beam path removed, to show linear translation stages and steering bars used to maintain alignment of X-ray source (A), analyser (B), and sample / detector assembly (C).

Fe K-edge XANES data were also acquired on the KMC-2 beamline at the BESSY-II synchrotron. The KMC-2 beamline is situated on a bending magnet and was configured with a Si (111) channel cut monochromator and mirror for harmonic rejection. Incident and transmitted beam intensities were measured using ionization chambers, filled with mixtures of He and N_2, operated in a stable region of their I/V curve. For synchrotron data, a step size of 0.25 eV was used in the XANES region with a count time of 1s / point. Data integration time was typically 0.5h per spectrum.

Samples were prepared from finely ground specimens of glasses or reference compounds dispersed in 70 mg of polyethylene glycol, pressed into 13 mm diameter pellets, to achieve a thickness of one absorption length. Data reduction and analysis was performed using the programs Athena and Hephaestus [42]. Calibration of the energy scale were achieved by reference to a 5 μm Fe foil (Goodfellow Cambridge Ltd.), for which the first peak in the derivative spectrum was set to 7112.00 eV, as defined by Bearden and Burr for the Fe K absorption edge [43]. Data from the reference foil were acquired periodically for the laboratory and synchrotron experiments, the absolute energy scale for the laboratory XANES data was calibrated as previously described [30, 44].

Extraction and analysis of the pre-edge feature in Fe K-edge XANES data were achieved using Microsoft Excel software, to first fit a spline function to model and subtract the contribution of the edge step. The pre-edge envelope was then deconvoluted by fitting Gaussian components using a linear least squares refinement to optimise the normalised height, full width at half height, and energy position, so as to minimise the difference between the observed and calculated envelope determined from the sum of the fitted Gaussian components. Fitting initially utilised two Gaussian components with the significance of a third component tested by inspection of the improvement in goodness of fit. The total integrated intensity and centroid energy of the pre-edge feature were determined, respectively, from the sum of the integrated intensities and intensity weighted average energy position of the Gaussian components. The same initial parameters were used for all data sets, and several sets of different initial parameters trialled, to ensure that convergence did not arise in a local minimum. This approach is a simplification of that applied by Wilke *et al.*, which employed pseudo-Voigt components [28, 30].

3. Results and discussion

Figure 2 shows a head to head comparison of Fe K-edge XANES data acquired from $40Fe_2O_3$–$60P_2O_5$ (mol%) glasses, where, respectively, C and M are used to denote materials processed by conventional and microwave heating, and PO and AP denote the use of P_2O_5 and $NH_4H_2PO_4$ reagents. Excellent correspondence is observed between the laboratory (points) and synchrotron (solid line) XANES data, demonstrating that acquisition of good quality Fe K-edge XANES data using the laboratory set up, with reasonable data integration times. Note that the synchrotron data have *not* been convoluted by Gaussian or Lorentzian broadening for this comparison.

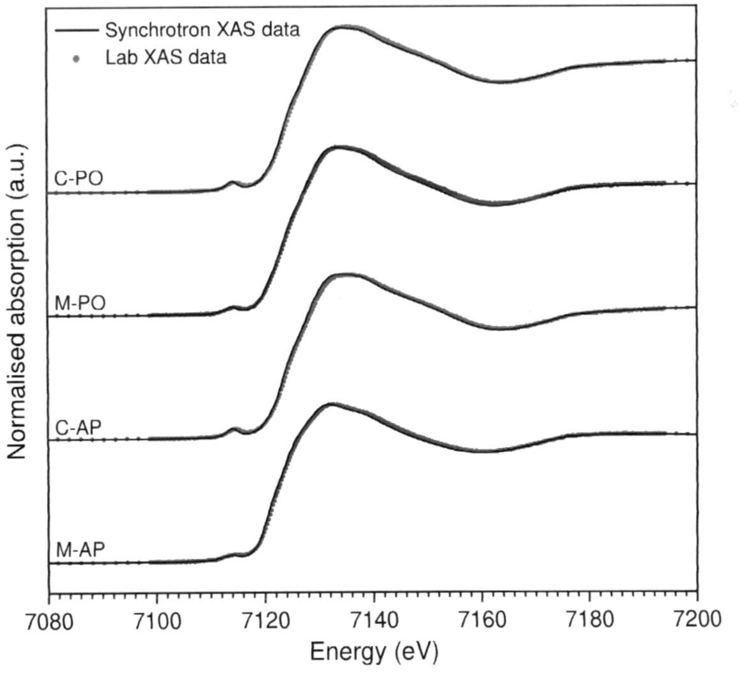

Figure 2. Comparison of normalised transmission mode Fe K-edge XANES spectra from $40Fe_2O_3$–$60P_2O_5$ glasses prepared by microwave (M) or conventional (C) heating of Fe_3O_4 with P_2O_5 (PO) or $NH_4H_2PO_4$ (AP), acquired using laboratory XAS spectrometer (points) and KMC-2 synchrotron beamline (solid line).

Figure 3 shows the extracted pre-edge features from the laboratory XANES data of the glass samples and the modelled envelope of the Gaussian components. Although the laboratory data show some scatter, as a result of the relatively low signal to noise ratio, the pre-edge features are evidently resolvable and measurable relative to the background. For each material, the pre-edge features of the laboratory and synchrotron XANES data were adequately modelled using three Gaussian components, the fitted parameters are summarised in Table 1. The fitted centroid position and integrated intensity of the pre-edge features are plotted in Figure 4, together with data from the reference compounds analysed in this study and that of Wilke *et al.* [28]. Note that the data of Wilke *et al.* (calibrated to E_0 = 7111.08 eV), were shifted by +0.92 eV, consistent with our absolute energy scale.

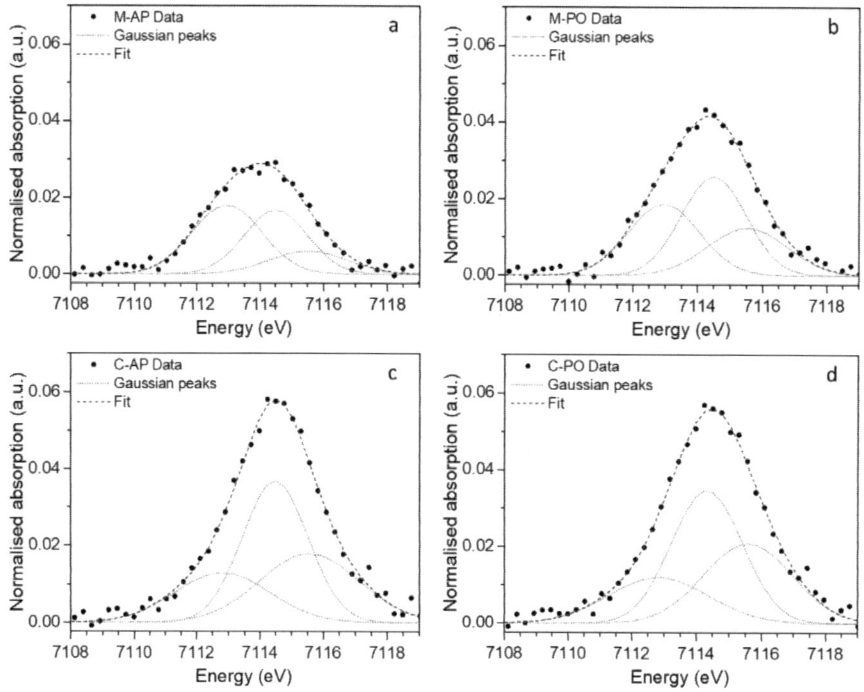

Figure 3. Normalised pre-edge spectra at Fe K edge (points) from $40Fe_2O_3$–$60P_2O_5$ glasses prepared by microwave (M) or conventional (C) heating of Fe_3O_4 with P_2O_5 (PO) or $NH_4H_2PO_4$ (AP), acquired using laboratory XAS spectrometer (points), showing fitted Gaussian components (grey dotted line) and their summation envelope (blue dotted line).

Table 1. Integrated intensity and centroid energy for Gaussian components fitted to pre-edge features of $40Fe_2O_3$–$60P_2O_5$ glasses prepared by microwave (M) or conventional (C) heating of Fe_3O_4 with P_2O_5 (PO) or $NH_4H_2PO_4$ (AP). Values in bold type are derived from laboratory data; values in normal type are derived from synchrotron data.

Sample	Gaussian 1		Gaussian 2		Gaussian 3		Centroid Energy (eV)	Total integrated intensity (a.u)
	Energy (eV)	Intensity (a.u.)	Energy (eV)	Intensity (a.u.)	Energy (eV)	Intensity (a.u.)		
C-AP	**7112.8**	**0.049**	**7114.5**	**0.096**	**7115.5**	**0.070**	**7114.4**	**0.215**
	7112.9	0.046	7114.4	0.103	7115.4	0.068	7114.4	0.217
C-PO	**7112.8**	**0.047**	**7114.3**	**0.095**	**7115.6**	**0.072**	**7114.4**	**0.213**
	7112.8	0.049	7114.3	0.120	7115.5	0.043	7114.3	0.212
M-AP	**7113.0**	**0.049**	**7114.5**	**0.042**	**7115.5**	**0.018**	**7114.0**	**0.110**
	7112.4	0.041	7114.4	0.073	7115.5	0.002	7113.7	0.116
M-PO	**7113.0**	**0.051**	**7114.5**	**0.065**	**7115.5**	**0.038**	**7114.2**	**0.154**
	7112.8	0.045	7114.3	0.094	7115.5	0.021	7114.1	0.160

As shown in the speciation plot of Figure 4, the data points for the Fe^{3+} reference compounds $NaFeSi_2O_6$ (octahedral, FeO_6) and $FePO_4$ (tetrahedral, FeO_4) are in excellent agreement between the laboratory data (solid blue circles) and synchrotron data (solid red circles), and the data reported by Wilke *et al.* (solid black circles) [28]. The Fe^{2+} reference compounds $Fe_{1.5}Mg_{0.5}Al_{9.1}Si_{3.9}(OH)_2$ (octahedral, FeO_6) and Fe_2SiO_4 (tetrahedral, FeO_4) were investigated using only the laboratory set up, but these data (solid blue circles) but are also in reasonable agreement with the data reported by Wilke *et al.* (solid black circles) [28]. Both the laboratory and synchrotron data points fall within the known field for Fe^{2+} / Fe^{3+} speciation for octahedral (CN = 6) and tetrahedral (CN = 4) co-ordination, defined by the reference compounds reported by Wilke *et al.* (solid / open black circles). Figure 4 demonstrates that laboratory XANES is effective in distinguishing Fe^{2+} and Fe^{3+} species according to the centroid of the pre-edge features at 7112.9 ± 0.1 eV and 7114.5 ± 0.1 eV respectively. The chemical shift between the centroid energy of Fe^{2+} and Fe^{3+} species determined in this investigation was thus 1.6 ± 0.1 eV for laboratory XANES data, which is in excellent agreement with the value of 1.4 ± 0.1 eV determined by Wilke *et al.* and our previous feasibility study [28, 30]. Figure 4 demonstrates that laboratory XANES is also effective in differentiating tetrahedral and octahedral Fe^{2+} / Fe^{3+} species, according to the relative total integrated intensity of the pre-edge feature. As noted above, the pre-edge feature is associated with 1s – 3d quadrupole transitions, which are only weakly allowed for a centrosymmetric octahedral environment but have enhanced transition probability in non-centrosymmetric tetrahedral environments, due to admixture of unoccupied 4p orbitals in the final state.

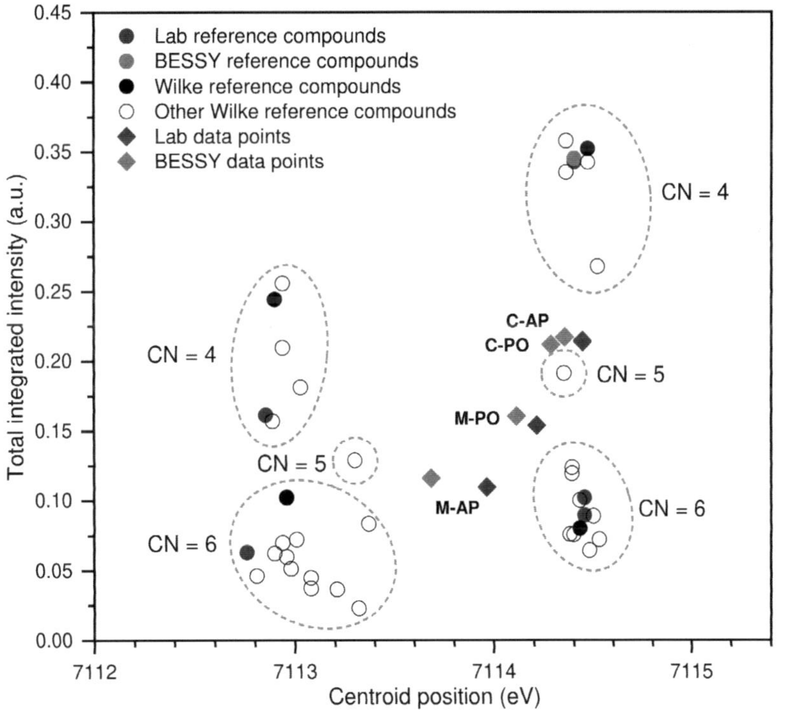

Figure 4. Speciation field diagram, based on total integrated intensity and centroid energy of pre-edge feature of Fe K-edge XANES data. Data points derived from analysis of our laboratory XANES are shown by solid blue circles (reference compounds) or diamonds (glasses); data points derived from synchrotron XANES are shown as solid red circles (reference compounds) or diamonds (glasses); equivalent data points of reference compounds from Wilke *et al.*, are shown as solid black circles (other reference data from Wilke *et al.*, shown as open black circles) [28]. Note data from Wilke *et al.*, are shifted by +0.92 eV consistent with our absolute energy scale as described in the text.

Turning to the analysis of the $40Fe_2O_3$–$60P_2O_5$ glass compositions, Figure 4 shows reasonably good agreement between the laboratory and synchrotron data points, demonstrating that laboratory XANES has, in principle, sufficient resolution and sensitivity to probe Fe^{2+} / Σ Fe ratio in iron phosphate glasses. The centroid energy of the pre-edge feature of glasses produced by microwave heating (M) is displaced toward lower energy compared to glasses produced by conventional melting (C), implying the presence of a greater concentration of Fe^{2+} species in the microwave processed materials. This is in agreement with our previous ^{57}Fe Mossbauer investigation [23], which estimated the Fe^{2+} / Σ Fe ratio to be 0.43 for glass M-AP, 0.22 for glass M-PO, and 0.10 for glass C-AP and C-PO. The centroid position of the pre-edge feature is known to show a non-linear dependence on Fe^{2+} / Σ Fe ratio, when both average oxidation state and co-ordination number vary at the same time [28], which is clearly inferred in this case from Figure 4 (see below). Hence, the Fe^{2+} / Σ Fe ratio cannot yet be reliably estimated from our XANES data without further careful calibration using mixtures of reference compounds. Nevertheless, it is evident from Figure 4 that glasses produced by microwave heating are characterised by a higher Fe^{2+} / Σ Fe redox ratio compared to the counterparts prepared by conventional melting. The higher Fe^{2+} / Σ Fe ratio for the microwave processed glasses, previously demonstrated by ^{57}Fe Mossbauer spectroscopy [23], is a reflection of the rapid processing time, which does not allow the melt to reach equilibrium with the prevailing oxygen partial pressure, compared to the conventionally processed counterparts. It is also evident from Figure 4, that the highest Fe^{2+} / Σ Fe ratio is observed for the microwave processed glass using $NH_4H_2PO_4$ as the phosphate source, also in agreement with our earlier ^{57}Fe Mossbauer investigation [23]. Decomposition of $NH_4H_2PO_4$ during the rapid microwave heating evolves NH_3 gas which maintains a sufficiently reducing environment to assist incorporation of Fe^{2+} within the melt, potentially also inducing partial reduction of Fe^{3+}. It is notable that the synchrotron determined data points in Figure 4 (red diamonds) are displaced to slightly higher centroid energy, compared to those determined by laboratory analysis (blue diamonds) on the same glass materials. This may suggest a small systematic difference in Fe^{2+} / Σ Fe ratio determined by laboratory and synchrotron XANES, however, this should be considered with due caution given the expected non-linear dependence of centroid energy on Fe^{2+} / Σ Fe ratio, which must be determined for both data sets.

Figure 4 also reveals a measurable difference of average Fe co-ordination in the microwave and conventionally melted glasses. In the case of conventionally melted glasses, the integrated intensity of the pre-edge feature implies a mixture of tetrahedral and octahedral, and/or pentahedral, Fe^{3+} species are present. This is in agreement with more recent Fe K-edge synchrotron XANES and ^{57}Fe Mossbauer studies of iron phosphate glasses, which point to a mixture of tetrahedral and octahedral, and or pentahedral, Fe^{3+} species, when Fe^{2+} / Σ Fe < 0.2 [13, 45, 46]. The microwave processed glasses are characterised by pre-edge features of lower integrated intensity, compared to their conventionally processed counterparts. This implies that the higher Fe^{2+} / Σ Fe ratio of the microwave processed glasses is associated with an increased proportion of octahedral Fe^{2+} (which would be expected to prefer octahedral co-ordination on the basis of crystal field stabilisation energy). This observation is consistent with the current understanding of the structure of iron phosphate glasses, in which an increase in Fe^{2+} / Σ Fe ratio, results in conversion of network forming tetrahedral Fe^{3+} to network modifying octahedral Fe^{2+} polyhedra [13,45,46]. Further work to accurately estimate the Fe^{2+} / Σ Fe ratio using mixtures of reference compounds, should also allow the proportion of co-ordination environments to be determined from laboratory XANES data,

4. Conclusions

Analysis of the weak pre-edge features in Fe K-edge XANES data from a laboratory XAS spectrometer demonstrated microwave processed iron phosphate glasses to incorporate a higher fraction of Fe^{2+} species, compared to counterparts produced by conventional melting, consistent with our previous ^{57}Fe Mossbauer spectroscopy study. Laboratory and synchrotron Fe K-edge XANES data from the same suite of glass materials were found to be in excellent agreement. Accurate determination of Fe^{2+} / Σ Fe ratio by laboratory XANES analysis, will require careful calibration against centroid energy of the pre-edge features, due to the known non-linear dependence when both average oxidation state and co-ordination number vary at the same time. Nevertheless, this preliminary study has established that

IOP Conf. Series: Materials Science and Engineering **818** (2020) 012020 doi:10.1088/1757-899X/818/1/012020

laboratory XANES data are of sufficient resolution and sensitivity, in principle, to routinely probe Fe^{2+} / Σ Fe ratio in iron phosphate glasses and other materials.

5. References

[1] NDA 2016 *Nuclear Decommissioning Authority Strategy effective from April 2016*

[2] NDA/RWM 2016 *Geological Disposal Science and Technology Plan.*

[3] NDA 2019 *5-year R&D Plan 2019 to 2024*

[4] Hyatt N C and James M 2013 *Nucl. Eng. Int.* **2** 10-3

[5] Bingham P A, Hyatt N C and Hand R J 2012 *Glass Technol. Part A* **53** 83.

[6] Bingham P A, Hyatt N C, Hand R J and Forder S D 2013 *Glass Technol. Part A* **54** 1.

[7] Bingham P A, Hyatt N C, Hand R J and Wilding C R 2009 *Mat. Res. Soc. Symp. P.* **1124** 161.

[8] Heath P G, Corkhill C L, Stennett M C, Hand R J, Whales K M and Hyatt N C 2018 *J. Nucl. Mater.* **508** 203.

[9] Hyatt N C, Morgan S, Stennett M C, Scales C R, Deegan D, 2007 *Mat. Res. Soc. Symp. P.* **985** 293

[10] Hyatt N C, Schwarz R R, Bingham P A, Stennett M C, Corkhill C L, Heath P G, Hand R J, James M, Pearson A and Morgan S 2013 *J. Nucl. Mater.* **444** 186.

[11] Brow R K, Kim C W and Reis S T 2020 *Int. J. Appl. Glass Sci.* **11** 4.

[12] Day D E, Wu Z, Ray C S and Hrma P 1998 *J Non-Cryst. Solids* **241** 1.

[13] Joseph K, Stennett M C, Hyatt N C, Asuvathraman R, Dube C L, Gandy A S, Kutty K V G, Jolley K, Rao P R V and Smith R 2017 *J. Nucl Mater.* **494** 342.

[14] Kim C W and Day D E 2003 *J. Non-Cryst. Solids* **331** 20.

[15] Mesko M G, Day D E and Bunker B C 2000 *Waste Manag.* **20** 271-8.

[16] Marasinghe G K, Karabulut M, Ray C S, Day D E, Shumsky M G, Yelon W B, Booth C H, Allen P G and Shuh D K 1997 *J. Non-Cryst. Solids* **222** 144.

[17] Karabulut M, Marasinghe G K, Ray C S, Day D E, Waddil G D, Booth C H, Allen P G, Bucher J J, Caulder D L and Shuh D K 2002 *J. Non-Cryst. Solids* **306** 182.

[18] Ray C S, Fang X, Karabulut M, Marasinghe G K and Day D E 1999 *J. Non-Cryst. Solids* **249** 1.

[19] Fang X, Ray C S, Moguš-Milanković A and Day D E 2001 *J. Non-Cryst. Solids* **283** 162.

[20] Griscom D L, Merzbacher C I, Bibler N E, Imagwa H, Uchiyama S, Namiki A, Marasinghe G K, Mesko M and Karabulut M 1998 *Nucl. Instr. and Meth. In Phys. Res. B* **141** 600.

[21] Joseph K, Jolley K and Smith R 2015 *J. Non-Cryst. Solids*, **411** 137.

[22] Jolley K and Smith R 2016 *Nucl. Instr. Methods Phys. Res. Sec. B.* **374** 8.

[23] Forder S D, Bingham P A, McGann O J, Stennett M C and Hyatt N C 2013 *Hyperfine Interact.* **217** 83.

[24] Waychunas G A, Apted M J and Brown G E 1983 *Phys. Chem. Miner.* **10** 1.

[25] Bajit S, Sutton S R and Delaney J S 1994 *Geochim. Cosmochim. Acta* **58** 5209.

[26] Galoisy L, Calas G and Arrio M A 2001 *Chem. Geol.* **174** 307.

[27] Petit P E, Farges F, Wilke M and Solé V A 2001 *J. Synchrotron Radiat.* **8** 952.

[28] Wilke M, Farges F, Petit P E, Brown G E and Martin F 2001 *Am. Mineral.* **86** 714.

[29] Berry A J, O'Neill H S, Jayasuriya K D, Campbell S J and Foran G J 2003 *Am. Mineral.* **88** 967.

[30] Mottram L M, Cafferkey S, Mason A R, Oulton T, Sun S K, Bailey D J, Stennett M C and Hyatt N C 2020 *J. Geosci.,* **65** 27.

[31] Stennett M C and Hyatt N C 2009 *Mat. Res. Soc. Symp. P.* **1124** 147.

[32] Mayzan M Z H, Stennett M C, Hyatt N C and Hand R J 2014 *J. Nucl. Mater.* **454** 343.

[33] Arnold H 1986 *Z. Kristallogr. Cryst. Mater.* **177** 139.

[34] Baum E, Treutmann W, Behruzi M, Lottermoser W and Amthauer G 1988 *Z. Kristallogr.* **183** 273.

[35] Smyth J R 1975 *Am. Mineral.* **60** 1092.

[36] Hawthorne F C, Ungaretti L, Oberti R, Caucia F and Callegari A 1993 *Can. Mineral.* **31** 551.

[37] Seidler G T, Mortensen D R, Remesnik A J, Pacold J I, Ball N A, Barry N, Styczinski M and Hoidn O R *2014 Rev. Sci. Instrum.* **85** 113906.

[38] Seidler G T, Ditter A S, Ball N A and Remesnik A J 2016 *J. Phys. Conf. Ser.* **712** 012015.

[39] Mortensen D R, Seidler G T, Ditter A S and Glatzel P 2016 *J. Phys. Conf. Ser.* **712** 012036.
[40] Jahrman E P, Holden W M, Ditter A S, Mortensen D R, Seidler G T, Fister T T, Kozimor S A, Piper L F J, Rana J, Hyatt N C and Stennett M 2019 *Rev. Sci. Instrum.* **90** 024106.
[41] Mortensen D R and Seidler G T 2017 *J. Electron. Spectrosc. Relat. Phenom.* **215** 8.
[42] Ravel B and Newville M 2005 *J. Synchrotron Radiat.* **12** 537.
[43] Bearden J A and Burr A F 1967 *Rev. Mod. Phys.* **39** 125.
[44] Mottram L M, Dixon Wilkins M C, Blackburn L R, Oulton T, Stennett M C, Sun S K, Corkhill C L and Hyatt N C 2020 *MRS Adv.* **5** 27.
[45] Wang G, Wang Y and Jin B 1994 *Proc. SPIE* **2287** 214.
[46] Karabulut M, Marasinghe G K, Ray C S, Day D E, Waddill G D, Allen P G, Booth C H, Bucher J J, Caulder D L, Shuh D K, Grimsditch M, Saboungi M L 2000 *J. Mater. Res.* **15** 1972.

Acknowledgment

We are grateful for financial support from the Nuclear Decommissioning Authority and EPSRC under grant numbers EP/M026566/1, EP/S01019X/1, EP/N017870/1 and EP/R511754/1. This research utilised the HADES / MIDAS facility at The University of Sheffield established with financial support from EPSRC and BEIS, under grant EP/T011424/1. This research was undertaken, in part, at the KMC-2 beamline at the BESSY-II synchrotron radiation source.

IOP Conf. Series: Materials Science and Engineering **818** (2020) 012021 doi:10.1088/1757-899X/818/1/012021

Influence of thermal treatment on the disposability of spent ion exchange resins in a deep geological repository: a French case

B Frasca[1], L Griffault[1], J Lefevre[1], C Martin[1] and M A Romero[1]

[1] ANDRA 1/7 Rue Jean-Monnet, 92298 Châtenay-Malabry, Paris, France

Email: lise.griffault@andra.fr

Abstract. The filtration of contaminated aqueous waste streams from nuclear applications produces spent ion exchange resins (IER) which can be classified either as low level Waste (LLW) or as intermediate level Waste (ILW). For the purpose of the work conducted in the framework of the European THERAMIN project, studies considered spent IER waste form to be routed to the French deep geological disposal facility in the Callovo-Oxfordian formation (Cigéo). This form of waste is known to release hydrogen by radiolysis degradation, reactive species and complexing compounds. Today, for disposability in Cigéo, direct cementation of IER is the main immobilization process. This work aims to evaluate the application of the incineration-vitrification with plasma process, SHIVA, on a mixture of zeolites, diatoms, and spent IER in regards to the reference immobilization process by cementation. Production of an alumina-borosilicate type glass using SHIVA incineration-vitrification process was considered in order to examine the impact of this process for managing the specificities of this IER waste form with the additional advantage of minimizing the disposal volume. The potential benefits that thermal treatment can provide in the context of Cigéo for such spent IER waste form are evaluated based on the physical characterizations and leaching experiments carried out by CEA on the resulting vitrified product. The influence of thermal treatment on several characteristics in line with a generic list of Waste Acceptance Criteria (WAC) defined within the THERAMIN project have been qualitatively evaluated considering their potential influences on operational and long-term safety.

1. Introduction

In a nuclear facility, circulating water and the water released to the environment are filtered to eliminate radiological contamination [1]. This filtration process induces waste that could be composed of ion exchange resins (IER), zeolites and diatoms. The direct cementation of this waste is the main immobilization process for disposability in Cigéo (French deep geological disposal project in the Callovo-Oxfordian formation, Figure 1).

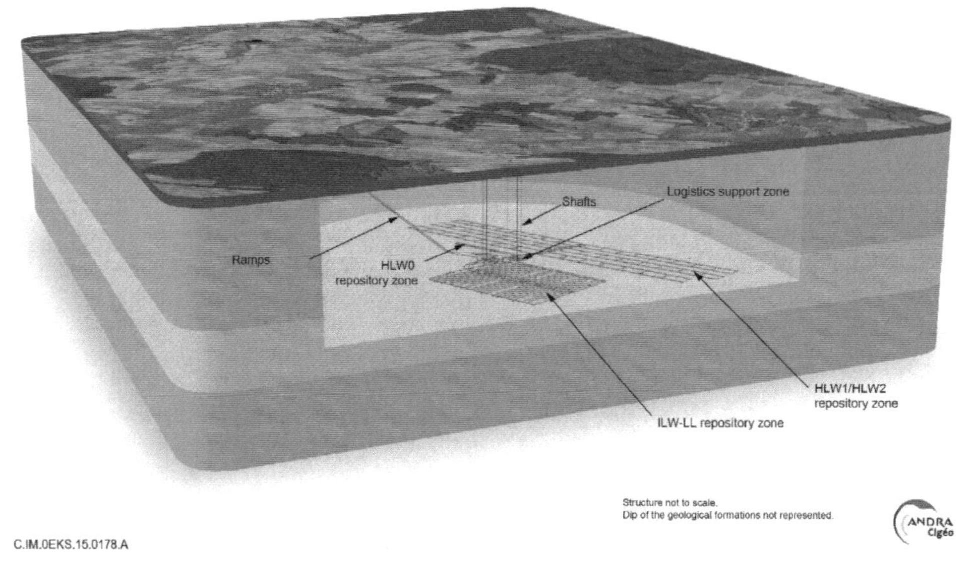

Figure 1. Schematic illustration of the French Cigéo project (extracted from [1]).

However, the weight ratio of such a process is around 10% of raw waste for 90% of cement, which significantly affects the disposal volume. Furthermore, as organic compounds, radiolysis of IER could lead to the production of hydrogen and complexing compounds. Therefore, vitrification of such waste could present an interesting alternative with a reduction of both volume and waste reactivity.

Spent IER waste forms in the French context may be categorised as Intermediate Level-Short Lived waste (ILW-SL), Low Level-Long Lived (LLW-LL) waste, or Intermediate Level-Long Lived (ILW-LL) waste. Considering the fact that in the French context, vitrified wastes are, so far, exclusively ILW-LL or HLW and therefore dedicated to deep geological disposal, this paper assesses the impact of vitrification on the disposability of a generic ILW-LL spent IER, mixed with zeolites and diatoms. This evaluation considers as baseline, a cementation process in concrete packaging.

In the framework of the THERAMIN project, an inactive mixture [2] composed of 45% of Zeolites, 44% of diatoms, 5.5% of anionic IER and 5.5% of cationic IER was vitrified with the inactive incineration-vitrification pilot process SHIVA developed by CEA [2][3]. The final product consists of an amorphous alumina-borosilicate glass (mainly composed of SiO_2, B_2O_3, Nd_2O_3 and Al_2O_3), which is homogeneous at the macroscopic level (Figure 2).

Figure 2. Waste glass sample from the SHIVA trial carried out in THERAMIN project [2].

Based on a list of generic WAC [4] to support analyses of the resulting product from thermal treatment, CEA performed physical characterizations and leaching experiments on the final product.

This assessment is carried out as part of WP4 of THERAMIN project. The article presents the main conclusions of this generic evaluation considering the main characteristics defined in the WP4 template, including the generic list of WAC.

2. Impact of thermal treatment on waste characteristics

2.1. Impact on radiological characteristics

The mixture used for SHIVA vitrification was inactive. Therefore, no specific activity measurement was performed and the characterization focused on the chemical composition before and after treatment.

Mixed waste produced by the filtration of contaminated aqueous waste streams may contain volatile radionuclides, such as ^{3}H, ^{14}C, ^{36}Cl, ^{106}Ru, ^{137}Cs and ^{129}I. A partial volatilisation of these elements is expected during the treatment. Most of them should be collected in the off -gas treatment system (which is not currently developed for active waste).

For a given waste volume, the SHIVA vitrification process leads to a higher concentration than the cemented IER waste form. From a qualitative analysis, it is likely that the concentration activity of the final vitrified waste form will increase.

Such evolution on radiological characteristics brings some elements to consider for the final product after thermal treatment:

- Heat generation and criticality: Even if the activity content of this type of IER waste does not usually cause criticality issues or heat generating problems, particular attention should be paid to activity concentration, heat production and fissile radioelements before and after processing due to volume reduction induced by the treatment.
- Waste classification: It should be noted that in France, the classification of radioactive waste is based on two important parameters to determine the appropriate management route and disposability: the level of waste activity and the half-lives of radionuclides. Depending on the impact of the thermal treatment on the concentration activity of the waste, the classification of the initial waste could be modified after thermal treatment (for example, a waste initially LLW-LL could be redirected to deep geological disposal if it becomes an ILW-LL waste). For the studied IER mixture, which is supposed categorised as an ILW-LL, there is no need to consider reorientation of disposability (i.e. Cigéo deep disposal).

2.2. Influence on the disposal volume

The baseline for this waste, classified as an ILW-LL, would be cementation with a usual weight ratio around 10% of raw waste. The mass distribution of waste and glass frit for SHIVA vitrification is respectively 40% and 60%. Considering a similar density of cemented waste and vitrified waste, the vitrification induces a volume smaller than cementation (by a factor of about 4). The completely burned IER will imply a larger factor (this will depend on the initial amount of IER).

In addition, the packaging increases this difference: cemented waste is usually packaged in concrete containers while vitrified waste is packaged in smaller metallic containers, which optimises the packaging volume and its subsequent disposal in the cell. In this evaluation, it is intended that the final product will be packaged in vitrified waste canister (CSD-V type).

The disposal of ILW-LL and HLW vitrified waste will require an over packing in the context of Cigéo. The primary packages delivered by producers will be introduced into specific standardised disposal packaging.

In the context of Cigéo, ILW-LL vitrified waste canisters would be packaged in a metallic over pack as for vitrified HLW (Figure 3). They would be disposed in the HLW area. One option would be to use them as spacers between HLW packages in order to limit the level of thermal output in the cell. Anyhow, for the purpose of the current analysis of the SHIVA final product, it is considered that it will be disposed in the HLW area with no need to consider its exact localisation (used as spacer or not).

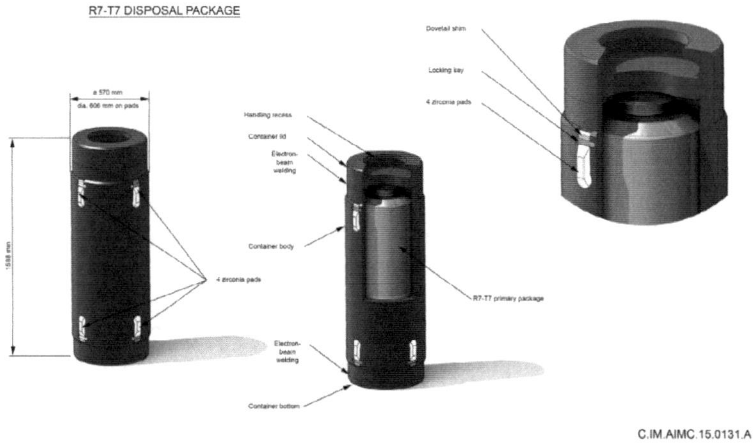

Figure 3. Schematic illustration of a vitrified high level waste package over pack (extracted from [5]).

2.3. Stability of the final glass product after thermal treatment

The volatilisation of radionuclides induces secondary waste from the off-gas treatment system. This volume has not yet been evaluated and will depend on the industrialization of the treatment process. However, with the SHIVA process, the gas produced during incineration of organic matter (around 1 m3 of gas per kg of organic matter) is burned [2], which leads to the absence of gas in the final glass form product.

This thermal treatment process will prevent H_2 generation in the waste package by radiolysis due to the degradation of organic matter and the absence of water in the glass waste form (unlike in the baseline IER cemented form). In addition, the elimination of organic compounds during thermal treatment will lead to the absence of organic complexants in the final waste form (unlike in the initial cemented form).

It is also noted that since the final product consists of an amorphous alumina-borosilicate glass, the release of radionuclides will depend on chemical durability of the vitreous matrix. In order to get preliminary information about chemical durability and alteration mechanisms of the final waste form, CEA performed some standard leaching tests [6].

The results of the leaching tests have been compared with those relating to an international reference glass of nuclear interest, called ISG, which includes the six main oxides of the typical French borosilicate glass SON68 used to vitrify high-level waste [7].

The comparison of the first leaching test results obtained with the SHIVA-VDM1 and ISG samples shows a similar trend [6]. The hydrolysis rate of the SHIVA-VDM1 sample (based on boron B release) is lower by approximately one order of magnitude than the hydrolysis rate of the ISG sample. Boron is considered as a tracer of glass alteration whereas silicon can be partially retained at the surface of the glass. This is clearly visible in the ISG results in Figure 4: Si release is lower than B release and indicates the possible formation of an alteration layer on the surface of the glass. This phenomenon could also explain why the alteration rate of ISG drops earlier than for the SHIVA-VDM1 sample. However, concluding about the long-term behaviour of the SHIVA-VDM1 waste form would require further investigation in the geochemical context of Cigéo.

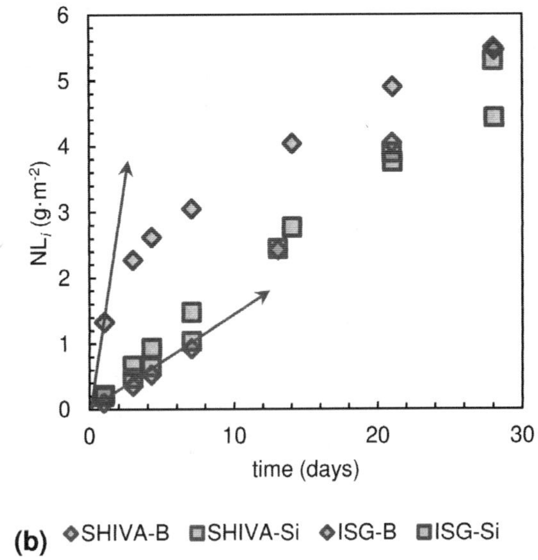

Figure 4. Compared evolutions of Normalized Leaching (NL) of Si and B during the leaching of ISG and SHIVA-VDM1 samples. The arrows are a guide for the eyes to estimate the hydrolysis rate of the matrices (extracted from [6]).

3. Impact of thermal treatment on disposability

3.1. Impact on disposability criteria

In France, Waste Acceptance Criteria (WAC) are defined together with the progressive development of Cigéo. Currently, only preliminary specifications are available. However, the glassy waste form is already taken into account in the Cigéo concept and so has associated preliminary WAC.

Among the list of generic WAC proposed in the framework of THERAMIN WP4 [4], some criteria have been considered according to the actual state of the Cigéo preliminary WAC. No prohibitive elements have been identified for the final borosilicate glass, with in addition the added benefit of the absence of H_2 release. The SHIVA vitrification process being tested on an inactive sample, the analysis is qualitative and could not be exhaustive at this point. However, considering the potential radionuclide inventory of IER, there should not be any heat generation due to concentration by volume reduction and no impact is expected on critical safety.

3.2. Impact on operational safety

The final product will be classified as an ILW-LL vitrified waste, which impacts on the packaging and overall disposal processes relative to the baseline spent IER waste.

The final product is intended to be packaged in the form of vitrified waste canister (CSD-V type), which is the main type of waste packaging defined for high-level vitrified waste to be disposed in Cigéo. The vitrified product obtained after thermal process of spent IER should therefore be disposed in the same conditions as other high-level waste [5].

However, due to a moderate initial activity of spent IER (classified as ILW), it is likely that high-level vitrified waste to be disposed in Cigéo present much higher values for (i) total activity per waste canister, (ii) mass of fissile material and (iii) heat generation. Therefore, the operational safety provisions that will be applied for disposal of high-level vitrified waste in Cigéo should cover operational safety to be considered for the final vitrified product obtained after thermal treatment of spent IER.

Vitrification resulting from the thermal treatment of the waste induced by the filtration of contaminated water (IER, zeolites, diatoms) offers, as would cementation, limitation to radioactive dispersion.

One advantage of thermal treatment for the operational safety compared to cementation is the absence of H_2 release, which simplifies the ventilation constraints in a disposal cell. On the other hand, due to the risk of re-concentration of the original inventory, in particular of fissile material, it will require confirmation that it is covered by the hypothesis retained for high-level vitrified waste.

3.3. Impact on long-term safety

The final product is an amorphous alumina-borosilicate glass macroscopically homogeneous in which radionuclides are integrated. From a qualitative point of view, analogies can be made with high-level vitrified waste already planned to be disposed in Cigéo [8].

The vitrified product obtained after thermal processing of spent IER will be disposed in the same conditions as other high-level vitrified waste (see Figure 3) and disposed in the high-level waste area of Cigéo. However, due to potential change of waste disposal area in Cigéo (from the ILW-LL area to HLW area), the redistribution of the inventory within the repository will have to be considered.

The glassy waste form produced after thermal treatment induces a slow release of radionuclides, which depends on the chemical durability of the vitreous matrix, compared to the instant release model associated to the baseline cemented spent IER waste form. This is more favourable even if long-term behaviour of the final product still needs to be assessed in more detail.

The long-term safety that will be applied for disposal of high level vitrified waste in Cigéo should cover long-term safety of the final vitrified product obtained after thermal treatment of spent IER. The overall long-term safety approach already taking into account high-level vitrified waste form can be applied (consideration of the same safety scenarios for example), only further knowledge on the chemical durability of the final product in the disposal conditions and new distribution of the inventory should be considered.

The absence of gas in the glass waste form after treatment leads to a reduction in the source term of gas after closure, which is favourable.

In addition, the characterisation results indicate that organic compounds have been removed leading to favourable characteristics for deep geological disposal due to the absence of organic complexants in the final product, although the Callovo-Oxfordian argillite offers good properties to limit radionuclide transfer.

4. Conclusion

This work, performed in the context of the European THERAMIN project, aimed at evaluating the application of the incineration-vitrification with plasma process, SHIVA, on a mixture of zeolites, diatoms, and IER, in regards to the reference immobilization process by cementation. A qualitative analysis was conducted in order to evaluate impacts of thermal treatment of a hypothetical l IER ILW-LL type of waste (baseline) into a vitreous product on disposability in Cigéo.

Results of the qualitative analysis indicate that the thermal treatment of a mixture of IER, zeolite and diatoms should induce less heterogeneous waste and a significant reduction of the waste and packaging volumes. The final product is an amorphous alumina-borosilicate glass with radionuclides integrated in a glassy matrix, which allows analogies with other vitrified waste to be disposed in the French deep geological disposal Cigéo.

Since the glassy waste form obtained with the SHIVA thermal treatment is already a type of waste taken into account in Cigéo, it is assumed that final glassy waste form should not require development of specific handling process and that operational safety applied for high level waste should cover the vitreous waste obtained by thermal treatment of spent IER.

From the generic list of THERAMIN WP4 WAC, qualitative analysis of some criteria according the actual state of the Cigéo preliminary WAC indicates no prohibitive elements for the final borosilicate glass with the added benefit of the absence of H2 release, thanks to the destruction of organic matter and the absence of water in the final waste form. This simplifies the ventilation constraints in disposal cell relative to the baseline IER cemented form.

Even if the activity content of this type of IER waste does not usually cause criticality issues or heat generation problems, particular attention should be paid to heat production and fissile radioelements before and after processing due to volume reduction induced by the treatment,.

Long-term safety already integrates such type of waste in Cigéo, which should not induce the development of specific approaches. In addition, the chemical durability of the glass appears more favourable than the instant release model associated to the baseline cemented IER waste form even if long-term behaviour of the final product needs to be assessed in more detail.

In addition, organic compounds have been removed which leads to favourable characteristics due to the absence of organic complexants although the Callovo-Oxfordian formation offers good properties which limit migration of radionuclides. Absence of gas in the glass waste form after treatment leads to a reduction in the source term of gas after closure, which is favourable.

Acknowledgements
This project has received funding from the Euratom research and training programme 2014-2018 under grant agreement No 755480. This paper reflects the authors' views and the European Commission is not responsible for any use that may be made of it.

References

[1] AIEA 1981 Management of spent-ion exchange resins from nuclear power plants *IAEA-TECDOC-238*

[2] CEA 2019 SHIVA and In-Can Melting technologies and demonstration test trials, *European Commission THERAMIN Deliverable D3.3.*

[3] Lemont F, Charvin P, Russello A and Poizot K 2010 An innovative hybrid process involving plasma in a cold crucible melter devoted to the future intermediate level waste treatment: The SHIVA technology, *In 12th International Conference on Modern Materials and Technologies (CIMTEC – 2010), Montecatini Terme, Italy*

[4] Andra *et al.* 2018 Waste Acceptance Criteria and requirements in terms of characterisation, *European Commission THERAMIN Deliverable D4.1.*

[5] Andra 2015a Safety Options Report – Operational Part (DOS-Expl). ANDRA report CG-TE-D-NTE-AMOA-SR2-0000- 15-0062, Paris (2015). Available from international.andra.fr.

[6] Andra *et al.* 2019 Characterization of thermally treated waste products, *European Commission THERAMIN Deliverable D4.2.*

[7] Gin S, Jolliver P, Fournier M, Berthon C, Wang Z, Mitroskov A, Zhu Z and Ryan JV 2015 The fate of silicon during glass corrosion under alkaline conditions: A mechanistic and kinetic study with the International Simple Glass, *Geochimica et Cosmochimica Acta,* **151**; pp. 68–85

[8] Andra 2015b Safety Options Report – Post-Closure Part (DOS-AF) ANDRA report CG-TE-D-NTE-AMOA-SR2-0000- 15-0062, Paris (2015). Available from international.andra.fr.

IOP Conf. Series: Materials Science and Engineering **818** (2020) 012022 doi:10.1088/1757-899X/818/1/012022

The HADES Facility for High Activity Decommissioning Engineering & Science: part of the UK National Nuclear User Facility

N C Hyatt[1]*, C L Corkhill[1], M C Stennett[1], R J Hand[1], L J Gardner[1] and C L Thorpe[1].

[1]Department of Materials Science & Engineering, The University of Sheffield, Mappin Street, Sheffield, S1 3JD.

*n.c.hyatt@sheffield.ac.uk

Abstract. Research and innovation is key to delivering UK Government's civil nuclear energy policy, in particular to accelerate reduction in the hazard, timescale and cost of legacy decommissioning and geological disposal of radioactive wastes. To address this challenge, a national centre of excellence, the HADES Facility, has been established to support research and innovation in High Activity Decommissioning Engineering & Science, as part of the wider network of UK National Nuclear User Facilities. Herein, we describe the development of this user facility, the current status of its capability, and functional equipment specifications. The unique capabilities of the HADES Facility, in the UK academic landscape, are emphasised, including: handling of weighable quantities of ^{99}Tc and transuranics; quantitative electron probe microanalysis of radioactive materials; hot isostatic pressing of radioactive materials; and laboratory-based X-ray absorption and emission spectroscopy. An example case study of the application of the HADES capability is described, involving thermal treatment of a real radioactive ion exchange resin waste to produce a conceptual vitrified waste form.

1. Introduction

The UK Government's civil nuclear policy has the fourfold objectives of: exploiting nuclear power for low carbon, reliable and affordable electricity generation; decommissioning of legacy nuclear fuel cycle activities; clean-up of defence nuclear liabilities; and safe disposal of radioactive wastes [1-7]. Building on the Nuclear Industrial Strategy, the Nuclear Sector Deal set the vision for the civil nuclear sector, to deliver by 2030 [3]:

- 30% cost reduction in cost of new build projects.
- Savings of 20% in cost of decommissioning compared with current estimates.
- Up to £2 billion domestic and international contract wins.

Research and innovation is recognised as crucial to meeting these challenges. In particular, there is a need for disruptive technologies to deliver safe and cost-effective management and disposal of radioactive wastes from legacy and future advanced fuel cycle operations. At the forefront of such technological development is thermal treatment of radioactive wastes, which utilises the application of heat to immobilise radionuclides and chemotoxic species within a passively safe product (a glass, slag or ceramic) suitable for interim storage and disposal. The benefits afforded by this approach include volume reduction (by removal of water, volatiles and voidage) and elimination of waste reactivity and organic inventory (by oxidation or thermal destruction). Nevertheless, technology maturation and implementation are critically dependent on [1, 8-10]: development of wasteform formulation and process envelopes; control of radionuclide volatility; understanding of product heterogeneity; and evaluation of wasteform behaviour in a Geological Disposal Facility (GDF). With regard to the latter, the Nuclear Decommissioning Authority (NDA) and its subsidiary, Radioactive Waste Management Ltd. (RWM), highlighted the need for research to understand the disposability of thermal treatment products, including those for plutonium disposition [1, 7-10].

NDA Strategy 3 prioritised the need to *develop higher activity waste strategy in a number of areas where there is the potential for beneficial change* [1], recognising that cement encapsulation was unlikely to be the optimal solution for all future waste streams. Accordingly, the new NDA 5-year R&D Plan 2019-24, prioritises as a Key R&D Topic – *Alternative waste treatment, with a particular focus on volume reduction understanding and, where appropriate, addressing the technical barriers to implementation of new thermal... treatment technologies across the NDA group* [10]. From a financial perspective, the potential reduction of 50% in packaged higher activity waste volume achieved by thermal treatment, could translate into UK-wide cost savings in intermediate level waste (ILW) packaging and storage costs, potentially of up to £1 billion [8].

Targeted estate-wide deployment of thermal treatment technologies could play a pivotal role in realising the 20% reduction in decommissioning costs demanded by HM Government's Nuclear Sector Deal [3]. The need for such intervention was recently highlighted by the House of Commons Public Accounts Committee [11, 12], who recommended NDA focus on delivering value for money to tax payers in decommissioning the Sellafield site and accelerating plutonium management strategy. To progress deployment of thermal treatment technology in UK decommissioning, NNL have invested in an active Geomelt^{TM} thermal treatment demonstrator on the Sellafield site, with inactive alternate technology demonstrator systems developed by the commercial sector, such as the Tetronics – Costain partnership to develop plasma vitrification technology. Whilst these pilot scale demonstrators have proven effective in building confidence in thermal processes and products, they rely on upstream small-scale laboratory research to design, develop and initially constrain formulation and process envelopes, and to gain early insight into downstream challenges such as off gas burden and product disposability. The window for insertion of thermal treatment technologies into lifecycle baseline plans is 2022-40 [8-9], and hence investment in enabling research and infrastructure is exceedingly timely.

We identified a national need for state of the art laboratory research infrastructure and equipment to ensure timely deployment of thermal treatment technology to meet the objectives of NDA Strategy and the Nuclear Sector Deal. To address this strategic need, the Engineering & Physical Science Research Council (EPSRC) and The University of Sheffield, announced, in spring 2020, the investment of £1M to establish the HADES Facility for High Activity Decommissioning Engineering & Science at The University of Sheffield, as part of the National Nuclear User Facility network. This leveraged prior recent investment of ca. £8M in existing state of the art infrastructure for radiological materials science, from UK Government, The University of Sheffield, EPSRC, NDA, and the Royal Academy of Engineering, and others.

2. Description of the HADES Facility

The HADES Facility is accommodated within ca. 400 m^2 of state of the art radiochemistry laboratories at The University of Sheffield. All laboratories are designated as Supervised Areas with relatively low hazard work with radionuclides permissible throughout the facility (kBq nat. U/Th). Additionally, there are Controlled Areas for more hazardous work with low inventories of radionuclides with high specific activity (100 kBq nat. U/Th; MBq ^{99}Tc; 10 kBq other β,γ). Investment of £1M by EPSRC and the University of Sheffield will upgrade a recently refurbished radiochemistry laboratory (60 m^2) to a Restricted Controlled Area, to satisfy demand for research using substantial inventories of radioactive materials, e.g. in a single procedure: 10 MBq of natural U/Th, or 5 MBq of mixed β,γ nuclides (e.g. ^{3}H, ^{14}C, ^{60}Co, ^{90}Sr, ^{137}Cs). Importantly, this will grow UK academic capacity and capability to meet demand for working with small quantities of real radioactive wastes, such as recent research on thermal treatment of spent ion exchange resins with an activity of 10 MBq kg^{-1} β,γ, contact dose rate 150 μSvh^{-1} – see case study in Section 4. In 2020, we will also commission a Restricted Controlled Area laboratory for working with transuranics within an integrated transuranic-materials glove box and furnace suite and a radiological Electron Probe Micro-Analyser (EPMA). These developments, funded by EPSRC as part of the Sheffield Hub of the Royce Institute, will deliver the first capability in the UK academic sector for working with weighable quantities of transuranic materials and accurate spatially-resolved microchemical analysis of radiological materials

3. HADES Facility Capability Platforms

Equipment and instrumentation within the HADES Facility is organised within the following capability platforms:

3.1. Materials handling

Negative pressure glove box workstations, dedicated for separate handling of U/Th and ^{99}Tc, are provided for safe manipulation of radiological materials under inert conditions (<ppm O_2, <ppm H_2O). For small radioactive inventories, which do not require handling under inert conditions, a suite of standalone containment glove boxes enable material manipulations under air or inert gas. Additionally, a standalone platform of three glove box workstations, with an integrated high temperature furnace (<1600°C) is dedicated to handling of transuranic materials.

3.2. Synthesis and processing

A suite of high temperature furnaces is available for the fabrication of glass, glass-ceramic and ceramic wasteforms, and investigation of thermal treatment processes, operating up to 1800°C under controlled atmosphere. Glass melts may be agitated by stirring or gas sparging, if required. Off gas analysis can be undertaken in real time by mass spectrometry or by chemical and radiochemical analysis of scrubber solutions. The HADES Facility also accommodates a unique capability in the UK for Hot Isostatic Pressing (HIP) of radiological materials using an AIP 6-30 HIP system with a custom built Active Furnace Isolation Chamber (AFIC); non-active and active materials may be processed up to 2000°C and 1350°C, respectively, under up to 200 MPa Ar. A suite of powder mixing, milling and grinding equipment is also available for high throughput batch preparation. A selection of non-stirred and stirred Parr pressure vessels are offered for hydrothermal and solvothermal syntheses.

3.3. Diffraction and spectroscopy

Rapid routine powder X-ray diffraction (XRD) analysis is undertaken using a Bruker D2 Phaser XRD system, with Ni filtered Cu Kα radiation and a Lynxeye energy discriminating detector. A Panalytical X'pert 3 XRD system, with a parabolic mirror for parallel beam Cu Kα radiation and PIXCel 1D silicon strip detector, is also available for high throughput data acquisition, grazing angle analysis, and *in situ* high temperature studies using an Anton Parr HTK 1200N furnace (up to 1200°C, under controlled atmosphere). Additionally, HADES is host to the UK's first laboratory end station for X-ray Absorption and Emission Spectroscopy (XAS / XES), which enables element-specific speciation studies in the range 4.5 – 18.0 keV, based on a customised EasyXAFS XES100 spectrometer, with ca. 1 eV instrument resolution achievable in the XANES region [13]. This unique capability is complemented by a Mossbauer spectrometer (Wissel MRG-500), for investigation of Fe and Sn speciation and local environment, including conversion electron detection for surface sensitive Mossbauer experiments (<100 nm for Fe). Confocal Raman spectroscopy with spatial and depth resolution, in the range 50-3200 cm^{-1} with 1.4 cm^{-1} spectral resolution, is enabled by an Horiba XPlora Plus system equipped with a 532 nm laser.

3.4. Microscopy and microanalysis

Imaging of microstructure and micro-chemical analysis is undertaken using a workhorse Hitatchi TM3030 scanning electron microscope (SEM), with magnification up to x30,000 in back scattered electron mode, equipped with an integrated Bruker Quantax 75 Energy Dispersive Spectrometer (EDS). The Hitachi TM3030 is also capable of imaging in low vacuum mode, for imaging of non-conductive, environmental or samples in their natural state. Access to a Philips XL30 SEM with an Oxford Instruments INCA EDX system is available for analysis of U/Th materials which require high magnification and/or topographical analysis in secondary electron mode. In 2020, as part of investment through the Sheffield Hub of the Royce Institute, a new JEOL JXA-8530F Electron Probe Micro Analysis system will be commissioned to enable quantitative microchemical analysis of radiological and nuclear materials by wavelength dispersive spectroscopy (WDS). 3D surface characterisation and imaging of materials is also possible using a Bruker Contour Elite K profilometer and Bruker MultiMode 8HR Atomic Force Microscope (AFM).

3.5. Thermal and physicochemical analysis
A Netzsch TG 449 F3 Jupiter simultaneous thermal analyser enables thermogravimetric and differential thermal analysis or scanning calorimetry (TGA and DTA / DSC) of materials, with coupled mass spectroscopy of evolved gases (up to 1500°C). For high throughput TGA-DSC analysis, a TA Instruments SDT 650 system is available (up to 1500°C). Additional physicochemical characterisation techniques include: glass rheology (Theta Rheotronic II; 5 x 10^{-3} to 5 x 10^{-6} Pa s, up to 1650°C); surface area analysis (Beckman Coulter SA-3100); helium pycnometry (Micromeritics Accupyc II); and particle size analysis (Malvern Mastersizer 3000).

3.6. Chemical and radiochemical analysis
A Thermo Fisher Inductively Coupled Plasma Optical Emission Spectrometer and Mass Spectrometer (ICP-OES 6000 iCAP and ICP-MS iCAP RQ) are available for trace and ultra-trace analysis of aqueous solutions, depending on analyte and concentration. Both systems are equipped with auto-samplers, and the ICP-MS system is equipped with an auto-dilution module, to enable rapid sample throughput. Two Thermo Fisher Ion Chromotography systems (ICS 1100) enable dissolved cation and anion analysis, with electrolytic suppression for low noise and drift, reduced background and lower detection limits; a coupled auto-sampler is available for high throughput analysis. These capabilities are complemented by a Hidex 300SL liquid scintillation counter using Triple to Double Coincidence Ratio (TDCR) technology for absolute activity measurement and separation of α and β scintillation. Accurate bulk chemical composition is determined using a Panalytical Zetium wavelength dispersive X-ray Fluoresence Spectrometer, equipped with a 4kW rhodium X-ray tube for high count rate and sample throughput and large area mapping capability (300 μm spot size); this was the first such state of the art system installed in the UK. In the next phase of Facility development, additional analytical capability will be procured including analysers for total organic carbon and oxygen / nitrogen determination.

3.7. Wasteform alteration and dissolution
A suite of temperature controlled ovens are dedicated to short and long term alteration studies of wasteform alteration behaviour using standardised and tailored protocols. Both batch and dynamic experiments are possible, including single pass flow through methodology for accurate determination of kinetic dissolution parameters. If required, experiments can also be performed under anoxic conditions, including periodic sampling.

3.8. Radiometrics and radiological protection
A Canberra BEGe high resolution gamma spectrometry system is available for identification and quantification of radionuclides in solid or liquid samples. In controlled areas, fixed personal contamination monitors are available, supplemented by a suite of portable large area survey meters, contamination monitors and dose rate detectors, and an Identifinder R100 for initial survey of received packages.

4. Case study of thermal treatment research enabled by the HADES Facility
In the UK context, spent ion exchange resins (SIERs) are classified as both intermediate and low level waste; some resins are considered to be *problematic radioactive waste... which has no defined waste treatment and disposal route available or for which existing routes are suboptimal* [14]. Challenges with SIERs include: incorporation of complexants, such as EDTA (ethylenediaminetetraacetic acid), and degradation of material in disposal environments to yield chelate molecules, which facilitate rapid

transport of complexed radionuclides; and degradation of material during storage to yield chemically toxic and volatile benzene, phenol, trietheylamine, and potentially ammonia gas [15-17].

One approach for effective conditioning of SIERs is thermal treatment utilising suitable glass forming additives to yield a passively safe glass or slag (crystallised glass) type wasteform [18-21]. We have investigated application of such an approach to support the strategic waste management considerations of a UK waste producer in partnership with Costain. In the present investigation the SIER waste is a mixture of Purolite NRW-100 strong acid cation and NRW-400 strong base anion resins, in a 1:1 chemical equivalent, with an admixture of EDTA complexant, see Figure 1. A full account of the research to demonstrate effective thermal treatment of this waste, at laboratory scale, will be published in due course. Here, we highlight some key outcomes of the second phase of the research programme, using real SIER waste, enabled by the capability of the HADES facility.

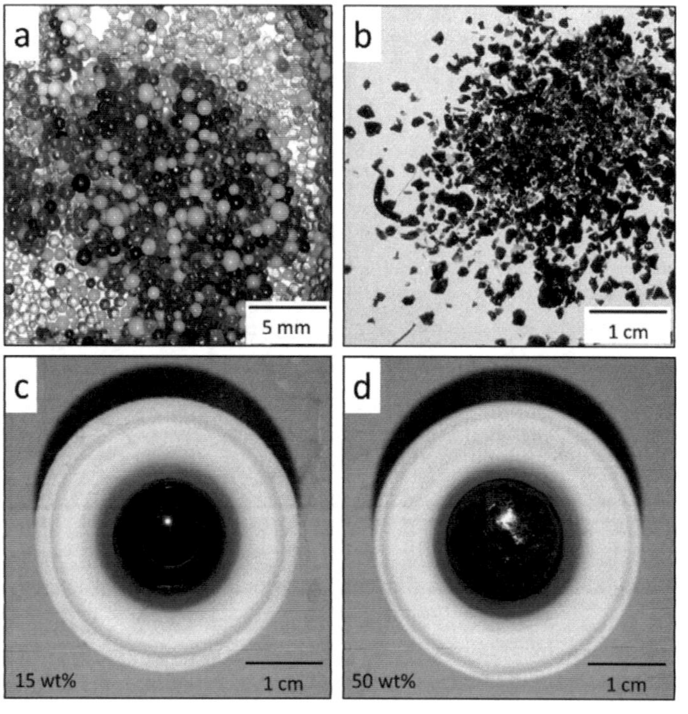

Figure 1/ a) Real spent ion exchange resin material; **b)** G73 glass frit; vitrified products produced from thermal treatment of **c)** 15wt% and **d)** 50wt% real spent ion exchange resin (on de-watered basis).

Table 1. Composition of G73 glass.

Component	Weight %
SiO_2	41.06
B_2O_3	2.06
Al_2O_3	0.28
Fe_2O_3	6.15
CaO	8.05
BaO	37.01
Li_2O	3.08

Our previous work had established the potential of barium borosilicate glasses for thermal treatment of SIERs, such as the G73 formulation summarised in Table 1, which has a high tolerance to sulphate formed by degradation of the resin sulphonate functionality [19-22]. Thermal treatment experiments at 10 – 40g scale were undertaken by first drying the resin to constant weight at 90°C and mixing with the required quantity of G73 glass frit in an alumina crucible, the mixture was ramped to 1000-1100°C at 3°C / min for 8h.

The first stage of our research showed that successful thermal treatment of inactive surrogate ion exchange resin at 1100°C, to yield a glass or slag product, could be achieved at an incorporation rate up to 50wt% of de-watered resin (20 wt% based on the dried resin equivalent), without separation of a sulphate salt. The real SIER material supplied for the purpose of research was expected to incorporate several inorganic radionuclides, e.g. ^{54}Mn, ^{55}Fe, ^{60}Co, ^{63}Ni, ^{125}Sb, ^{137}Cs. However, our high resolution γ-spectroscopy investigation revealed only the presence of ^{60}Co within the supplied material, in addition to ^{14}C and ^{3}H known to be present. The contact dose rate of the material was approximately 150 μSv h^{-1}, necessitating the use of lead shielding for extended manipulation procedures and storage of the material. To facilitate the active thermal treatment experiments, a purpose built vertical tube furnace was utilised, with a constant flow of air (120 or 240 ml min^{-1}) in order to carry through any volatilised ^{60}Co to be captured within a water scrubber or activated charcoal filter. However, no such release of ^{60}Co was detected from post-mortem γ-spectroscopy of scrubber liquor or filter material. Visual inspection and XRD analysis of thermal treatment experiments performed at 50wt% incorporation rate or below (120 ml min^{-1} air sparge), indicated that the process resulted in complete destruction of the resin feed, and afforded primarily a vitrified material with minor separation of a sulphate salt phase. SEM/EDX analysis of the vitrified material (Hitachi TM3030 instrument operating at 15kV), see Figure 2, revealed a homogeneous microstructure, with trace inclusions of an iron sulphide phase, which was expected given the high sulphur content and reducing nature of the melt. Further investigation demonstrated that an incorporation rate of 80wt% of dewatered resin could produce a slag like product with complete destruction of resin material, with an increased air sparge of 240 ml min^{-1}.

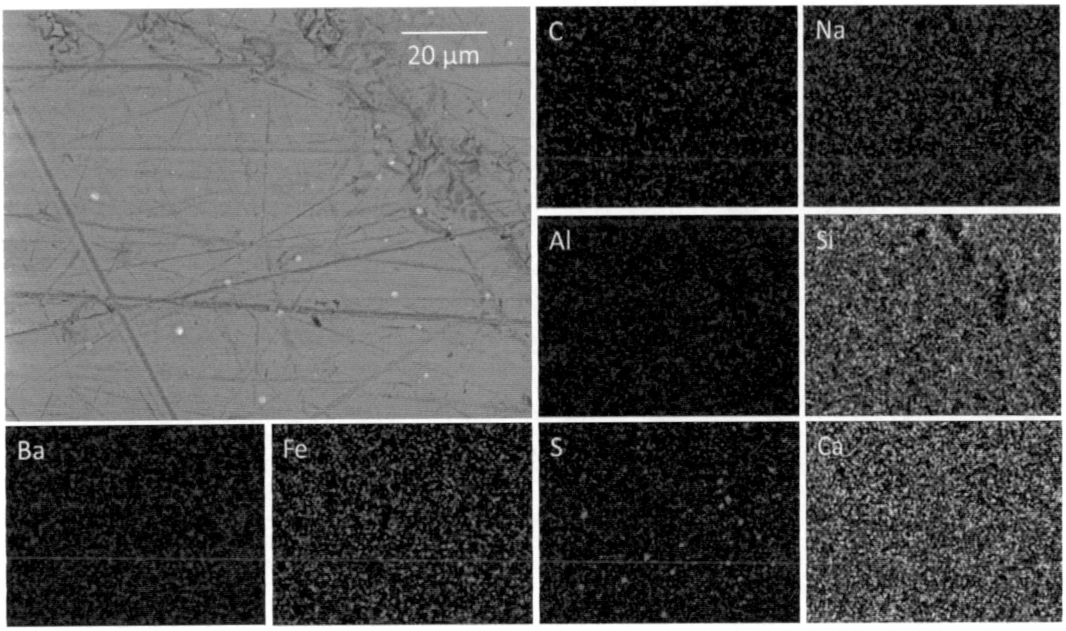

Figure 2. Backscattered electron image and elemental maps acquired from a representative area of the vitrified material obtained from thermal treatment of 50wt% real spent ion exchange resin (on de-watered basis) with G73 glass.

The process envelope evaluated at laboratory scale in this research is well within the application range of plasma vitrification technology, which is well suited to thermal treatment of organic radioactive wastes, such as resins and polymers [23, 24]. The conceptual products arising from thermal treatment of ion exchange resins are thought to meet the material related waste acceptance criteria of the Low Level Waste Repository. Therefore, the research summarised here has achieved the primary aim of demonstrating the efficacy of thermal treatment for resin destruction and production of a passively safe and homogeneous solid waste form, with minimal voidage, by destruction of the EDTA complexant and organic resin material. In particular, we emphasise the importance of validation of early stage surrogate research, using real radioactive waste resins, to develop early confidence in the compatibility of the approach.

5. Conclusions

The HADES Facility is established as a national centre of research excellence to support the UK nuclear decommissioning and disposal programme, in particular the development of thermal treatment technologies. Investment by The University of Sheffield, UK Government, and EPSRC has established state of the art capability and instrumentation for the handling of high inventories of radiological materials, including real radioactive wastes, to facilitate laboratory scale investigation of waste immobilisation technologies, characterisation of conceptual products, and evaluation of waste form disposability. The Facility provides unique capability within the UK academic landscape, including: handling of weighable quantities of ^{99}Tc and transuranics; quantitative electron probe microanalysis of radioactive materials; HIP of radioactive materials; laboratory-based X-ray absorption and emission spectroscopy. The HADES facility and equipment are open to external access via the arrangements of the UK National Nuclear User Facility and Royce Institute.

As shown in the case study of Section 4, the expertise and equipment established within the HADES Facility provide essential capability required to work with real radioactive wastes to develop and translate basic science and engineering into waste treatment processes, required to support strategic decision making. Such research capability is expected to play a pivotal role in accelerating the reduction in hazard, cost and timescale of UK nuclear decommissioning and enabling the case to be made for geological disposal of radioactive wastes, as set out in UK Government policy and the strategic plans of NDA and RWM.

6. References

[1] Nuclear Decommissioning Authority Strategy effective from April 2016, NDA, March 2016.
[2] UK Nuclear Innovation and Research Programme Recommendations, NIRAB, March 2016, NIRAB-75-10.
[3] Industrial Strategy, Nuclear Sector Deal, HM Government, 27 June, 2018.
[4] The Clean Growth Strategy Leading the way to a low carbon future, HM Government, October 2017, Amended April 2018.
[5] Nuclear Industrial Strategy - The UK's Nuclear Future, HM Government, 26 March 2013.
[6] Nuclear Liabilities Management Strategy, Ministry of Defence, September 2018.
[7] Geological Disposal Science and Technology Plan, RWM, May 2016, NDA/RWM/121.
[8] Review of the Integrated Project Team on Thermal Treatment and its R&D Programme, Issue 1, January 2019, NDARB034.
[9] Sellafield medium to long term research needs: Supporting the mission at Sellafield, Sellafield Ltd, 30 October 2018.
[10] NDA 5-year R&D Plan 2019 to 2024, NDA, 23 April 2019.
[11] Nuclear Decommissioning Authority: risk reduction at Sellafield, Sixty-Fifth Report of Session 2017–19, House of Commons Committee of Public Accounts, 31 October 2018, HC1375.
[12] Treasury Minutes: Government response to the Committee of Public Accounts on the Sixty-Fourth to the Sixty-Eighth reports from Session 2017-19, January 2019, CP 18.
[13] Mottram L M, Dixon Wilkins M C, Backburn L R, Oulton T, Stennett M C, Sun S K, Corkhill C L, and Hyatt N C 2020 *MRS Adv.* **5** 27.
[14] Problematic Waste Inventory Summary - May 2018, LLWR, NWP/REP/196, 2018.

[15] Wang J and Wan Z 2015 *Prog. Nucl. Energy* **78** 47.
[16] Rizzato C, Rizzo A, Heisbourg G, Večerník P, Bucur C, Comte J, Lebeau D, and Reiller P E, State of the art review on sample choice, analytical techniques and current knowledge of release from spent ion-exchange resins (D4.1), EC CAST Project, 2015.
[17] Geological Disposal, Carbon-14 Project Phase 2: Overview NDA/RWM/137, RWM, 2016.
[18] Hyatt N C and James M, 2013, *Nucl. Eng. Int.* **2** 10.
[19] Bingham P A, Hand R J, and Hyatt N C, 2012 *Glass Technol. Part A* **53** 83.
[20] Bingham P A, Hyatt N C, Hand R J, and Forder S D, 2013 *Glass Technol. Part A* **54** 1.
[21] Bingham P A, Hyatt N C, Hand R J, and Wilding C R, 2009 *Mat. Res. Soc. Symp. P.* **1124** 161.
[22] Heath P G, Corkhill C L, Stennett M C, Hand R J, Whales K M, and Hyatt N C, 2018 *J. Nucl. Mater.* **508** 203.
[23] Hyatt N C, Morgan S, Stennett M C, Scales C R, and Deegan D, 2006 *Mat. Res. Soc. Symp. P.* **985** 293.
[24] Hyatt N C, Schwarz R R, Bingham P A, Stennett M C, Corkhill C L, Heath P G, Hand R J, James M, Pearson A, and Morgan S, 2013 *J. Nucl. Mater.* **444** 186.

Acknowledgment

The authors gratefully acknowledge the University of Sheffield, Department of Business, Energy and Industrial Strategy, and the Engineering and Physical Sciences Research Council, for financial support to establish the HADES Facility, building on earlier investment through the MIDAS project; in particular under grant references EP/P02470X/1, EP/P025285/1, EP/N017617/1, EP/M026566/1, EP/L014041/1 and EP/T011424/1. We also acknowledge relevant EPSRC research support, to build the HADES research capability, under grants EP/N017617/1, EP/N017870/1, EP/M026566/1, EP/N017374/1 and EP/S012400/1. Research described in Section 4 was supported by the Submarine Development Agency and Costain, and we gratefully acknowledge the support of Mr. Leigh Wakefield and Dr. Christine Bruce in undertaking this research. The HADES Facility enabled research undertaken by the THERAMIN project which received funding from the Euratom research and training programme 2014-2018 under grant agreement No 755480.

AUTHOR INDEX